JUMP Math 5.2

Book 5 Part 2 of 2

Contents

jump math™

MULTIPLYING POTENTIAL.

JUMP Math
One Yonge Street, Suite 1014
Toronto, Ontario M5E 1E5
Canada
www.jumpmath.org

Writers: Dr. John Mighton, Dr. Anna Klebanov, Dr. Sohrab Rahbar, Sam Mercurio
Consultant: Dr. Sindi Sabourin
Editors: Megan Burns, Julia Cochrane, Janice Dyer, Laura Edlund, Ewa Krynski, Liane Tsui
Layout and Illustrations: Linh Lam, Gabriella Kerr, Laura Brady, Marijke Friesen,
                          Pam Lostracco, Ilyana Martinez
Cover Design: Blakeley Words+Pictures
Cover Photograph: © iStockphoto.com/Grafissimo

ISBN 978-1-927457-15-3

First printing November 2013

Printed and bound in Canada

A note to educators, parents, and everyone who believes that numeracy is as important as literacy for a fully functioning society

Welcome to JUMP Math

Entering the world of JUMP Math means believing that every child has the capacity to be fully numerate and to love math. Founder and mathematician John Mighton has used this premise to develop his innovative teaching method. The resulting resources isolate and describe concepts so clearly and incrementally that everyone can understand them.

JUMP Math is comprised of teacher's guides (which are the heart of our program), interactive whiteboard lessons, student assessment & practice books, evaluation materials, outreach programs, and teacher training. The Common Core Editions of our resources have been carefully designed to cover the Common Core State Standards. All of this is presented on the JUMP Math website: **www.jumpmath.org**.

Teacher's guides are available on the website for free use. Read the introduction to the teacher's guides before you begin using these resources. This will ensure that you understand both the philosophy and the methodology of JUMP Math. The assessment & practice books are designed for use by students, with adult guidance. Each student will have unique needs and it is important to provide the student with the appropriate support and encouragement as he or she works through the material.

Allow students to discover the concepts by themselves as much as possible. Mathematical discoveries can be made in small, incremental steps. The discovery of a new step is like untangling the parts of a puzzle. It is exciting and rewarding.

Children will need to answer the questions marked with a ▯ in a notebook. Grid paper notebooks should always be on hand for answering extra questions or when additional room for calculation is needed.

Contents

Unit 4: Number and Operations in Base Ten: Division

Unit 5: Number and Operations—Fractions: Fractions

Unit 6: Number and Operations in Base Ten: Decimals

Unit 3: Operations and Algebraic Thinking: Expressions and Equations

Unit 4: Number and Operations in Base Ten: Multiplying and Dividing Decimals

Unit 5: Measurement and Data: US Customary Units

Unit 6: Measurement and Data: Area and Volume

Unit 7: Geometry: 2-D Shapes

G5-1 Columns and Rows

1. Join the dots in the given column, row, or both.

   a) Column 2

   3 • | •
   2 • | •
   1 • | •
      1 2 3

   b) Row 3

   3 • • •
   2 • • •
   1 • • •
      1 2 3

   c) Row 1

   3 • • •
   2 • • •
   1 • • •
      1 2 3

   d) Column 1

   3 • • •
   2 • • •
   1 • • •
      1 2 3

   e) Column 2, Row 1

   3 •—• •
   2 • | •
   1 • | •
      1 2 3

   f) Column 2, Row 3

   3 • • •
   2 • • •
   1 • • •
      1 2 3

   g) Column 3, Row 1

   3 • • •
   2 • • •
   1 • • •
      1 2 3

   h) Column 1, Row 2

   3 • • •
   2 • • •
   1 • • •
      1 2 3

2. Circle the dot in the given position.

   a) Column 2, Row 1

   3 • • •
   2 • • •
   1 • ⊙ •
      1 2 3

   b) Column 3, Row 2

   3 • • •
   2 • • •
   1 • • •
      1 2 3

   c) Column 3, Row 1

   3 • • •
   2 • • •
   1 • • •
      1 2 3

   d) Column 2, Row 2

   3 • • •
   2 • • •
   1 • • •
      1 2 3

3. Circle the dot where the two lines meet. Then identify that dot's column and row.

   a) 3 • • •
       2 • | •
       1 •——•——•
          1 2 3

       Column _____

       Row _____

   b) 3 • • •
       2 •——• •
       1 • • •
          1 2 3

       Column _____

       Row _____

   c) 3 •——• •
       2 • • •
       1 • • •
          1 2 3

       Column _____

       Row _____

   d) 3 •——•——•
       2 • | •
       1 • • •
          1 2 3

       Column _____

       Row _____

4. Identify the column and the row for the circled dot.

   a) 3 • ⊙ •
       2 • • •
       1 • • •
          1 2 3

       Column _____

       Row _____

   b) 3 • • •
       2 • • ⊙
       1 • • •
          1 2 3

       Column _____

       Row _____

   c) 3 • • •
       2 ⊙ • •
       1 • • •
          1 2 3

       Column _____

       Row _____

   d) 3 • • •
       2 • • •
       1 • • ⊙
          1 2 3

       Column _____

       Row _____

You can write the column and the row for a point in brackets.
Always write the column first.

(5, 3)

column ↗ ↖ row

5. Circle the dot in the given position.

a) (2, 1)

```
3  •  •  •
2  •  •  •
1  •  •  •
   1  2  3
```

b) (3, 3)

```
3  •  •  •
2  •  •  •
1  •  •  •
   1  2  3
```

c) (1, 2)

```
3  •  •  •
2  •  •  •
1  •  •  •
   1  2  3
```

d) (2, 3)

```
3  •  •  •
2  •  •  •
1  •  •  •
   1  2  3
```

e) (3, 1)

```
3  •  •  •
2  •  •  •
1  •  •  •
   1  2  3
```

f) (3, 2)

```
3  •  •  •
2  •  •  •
1  •  •  •
   1  2  3
```

g) (1, 3)

```
3  •  •  •
2  •  •  •
1  •  •  •
   1  2  3
```

h) (2, 2)

```
3  •  •  •
2  •  •  •
1  •  •  •
   1  2  3
```

You can use letters instead of numbers to label columns and rows.

6. Circle the given point.

a) (A, 3)

```
3  •  •  •
2  •  •  •
1  •  •  •
   A  B  C
```

b) (Y, B)

```
C  •  •  •
B  •  •  •
A  •  •  •
   X  Y  Z
```

c) (0, 2)

```
3  •  •  •
2  •  •  •
1  •  •  •
   0  1  2
```

d) (0, 0)

```
2  •  •  •
1  •  •  •
0  •  •  •
   0  1  2
```

e) (A, C)

```
D  •  •  •  •
C  •  •  •  •
B  •  •  •  •
A  •  •  •  •
   A  B  C  D
```

f) (2, X)

```
Z  •  •  •  •
Y  •  •  •  •
X  •  •  •  •
W  •  •  •  •
   1  2  3  4
```

g) (4, 1)

```
4  •  •  •  •
3  •  •  •  •
2  •  •  •  •
1  •  •  •  •
   1  2  3  4
```

h) (3, 4)

```
4  •  •  •  •
3  •  •  •  •
2  •  •  •  •
1  •  •  •  •
   1  2  3  4
```

7. Write the position of the circled dot. Remember: the column is always given first.

a)
```
3  •  ⊙  •
2  •  •  •
1  •  •  •
   1  2  3
```
(_____, _____)

b)
```
3  •  •  •
2  •  •  ⊙
1  •  •  •
   1  2  3
```
(_____, _____)

c)
```
3  •  •  •
2  ⊙  •  •
1  •  •  •
   1  2  3
```
(_____, _____)

d)
```
3  •  •  •
2  •  •  •
1  •  •  ⊙
   1  2  3
```
(_____, _____)

G5-2 Coordinate Grids

The numbers in brackets that give the position of a point on a grid are called the **coordinates** of the point. They are also called an **ordered pair**.

1. a) Plot and label the points on the coordinate grid.
Cross out the coordinates as you go.

A̶ ̶(̶1̶,̶ ̶5̶)̶   B (1, 7)   C (3, 7)

D (6, 4)   E (7, 4)   F (8, 3)

G (7, 3)   H (5, 1)   I (5, 0)

J (4, 1)   K (4, 2)

b) Join the points in alphabetical order. Then join A to K.

c) What does the picture you made look like?

_____

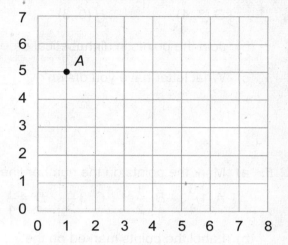

2. Write the coordinates of each point.

A (,   )   B (,   )

C (,   )   D (,   )

E (,   )   F (,   )

G (,   )   H (,   )

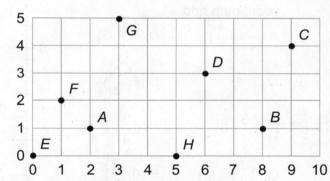

We use number lines to mark the grid lines.
The number lines are called **axes**.
One number line is called an **axis**.
The axes meet at the point (0, 0), called the **origin**.

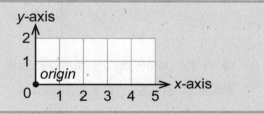

3. a) Fill in the coordinates for the given points.

A (1 , 3)   B (,   )   C (,   )

D (,   )   E (,   )   F (,   )

G (,   )   H (,   )   I (,   )

J (,   )   K (,   )   L (,   )

b) Which points are on the *x*-axis? _____

c) Which points are on the *y*-axis? _____

d) Which point is the origin? _____

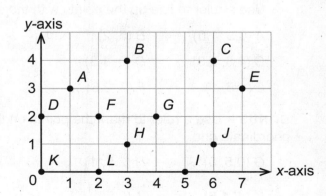

4. a) Plot and label the points on the coordinate grid.
Cross out the coordinates as you plot them.

A (0, 4)   B (4, 5)   C (1, 1)

D (4, 2)   E (4, 1)   F (0, 0)

G (3, 4)   H (0, 3)

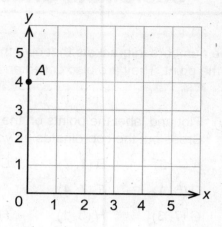

b) Join the points in alphabetical order. Then join A to H.

c) What letter have you drawn? _____

5. a) Mark the points on the number line.

$A \ 1\frac{3}{4}$ $B \ \frac{1}{2}$ $C \ 3\frac{1}{2}$ $D \ 4\frac{1}{4}$

b) Label the points marked on the coordinate grid.

$A \ (\frac{1}{4}, 1\frac{1}{2})$ $B \ (4\frac{1}{2}, 3)$

$C \ (2, 1)$ $D \ (6, \frac{3}{4})$

$E \ (2\frac{3}{4}, 0)$ $F \ (6\frac{3}{4}, 2\frac{1}{4})$

$G \ (3\frac{1}{4}, 3\frac{1}{4})$ $H \ (0, \frac{3}{8})$

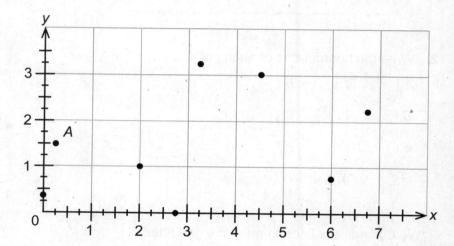

6. a) Mark the points on the number line.

A 1.8   B 0.5   C 3.1   D 4.8

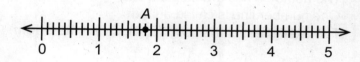

b) Label the points marked on the coordinate grid.
Use a ruler to line up the points with the axes.

A (0.5, 1.5)   B (5, 2)

C (4.8, 0.3)   D (0, 1.3)

E (1.5, 1)   F (3, 2.6)

BONUS ▶ Use a ruler to mark the points on the coordinate grid.

G (0.5, 0)   H (2.5, 1)

$I \ (4, 1.5)$ $J \ (3.5, \frac{1}{2})$

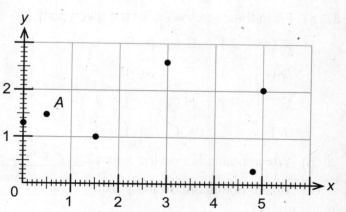

G5-3 Sliding Points

Josh **slides** a dot from one position to another. To move the dot from position 1 to position 2, Josh slides the dot **4 units right**.

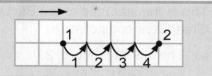

1. How many units **right** did the dot slide from position 1 to position 2?

   a)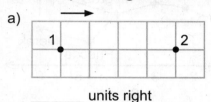

   _____ units right

   b)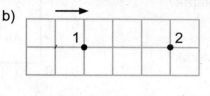

   _____

   c)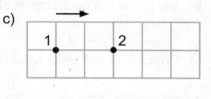

   _____

2. How many units **left** did the dot slide from position 1 to position 2?

   a)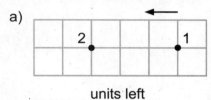

   _____ units left

   b)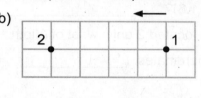

   _____

   c)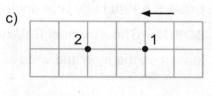

   _____

3. Follow the instructions to slide the dot to a new position.

   a) 3 units right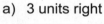

   b) 4 units left

   c) 5 units right

   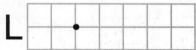   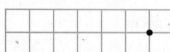   

4. How many units **right** and how many units **down** did the dot slide from position 1 to position 2?

   a)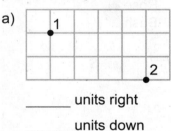

   _____ units right

   _____ units down

   b)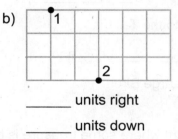

   _____ units right

   _____ units down

   c)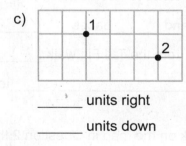

   _____ units right

   _____ units down

5. Slide the dot.

   a) 5 units right; 2 units down

   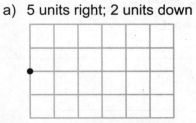

   b) 4 units left; 2 units up

   c) 3 units left; 4 units down

   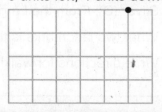

G5-4 Maps

1. This is the star map of the Big Dipper, part of the Great Bear constellation.

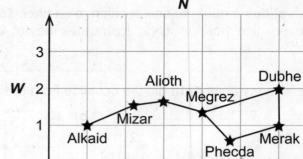

   a) The star at point (6, 2) is the official star of Utah.

   What is it called? _____

   b) What are the coordinates of Merak?

   (_____, _____)

   c) Which star is 5 units west of Merak?

   _____

   d) Galaxy M81 is located 1 unit north and 1 unit east from Dubhe. Mark it on the map.

   BONUS ▶ What star is at point (3, 1.6)? _____

   BONUS ▶ The Pinwheel Galaxy is located 2 units west of Alioth.

   Mark it on the map and write its coordinates. (_____, _____)

2. This map shows part of Treasure Island, where pirates have buried gold, silver, and weapons. Fill in the directions.

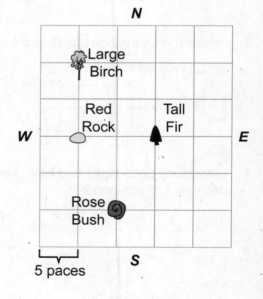

   a) From the Tall Fir, walk ___10___ paces (steps) ___west___ to the Red Rock.

   b) From the Red Rock, walk _____ paces north to the Large Birch.

   c) From the Red Rock, walk _____ paces _____

   and _____ paces east to the Rose Bush.

   d) From the Rose Bush, walk _____ paces _____

   and _____ paces _____ to the Tall Fir.

   e) From the Tall Fir, walk _____ and

   _____ to the Large Birch.

3. Mark on the map in Question 2 the point where some treasure is buried.

   a) Gold (G): From the Tall Fir, walk 5 paces east and 10 paces north.

   Weapons (W): From the Rose Bush, walk 10 paces west and 5 paces south.

   Silver (S): From the Large Birch, walk 10 paces south and 5 paces east.

   b) What two landmarks is the silver buried between? _____

   c) Write directions for walking from Gold to Silver.

   _____

4. This map shows all of Treasure Island. Each square on the map has sides 1 km long.

a) Round Lake is at point (2, 4). What is at point …

(3, 2)? _____

(3, 5)? _____

(5, 5)? _____

BONUS ▶ (6.5, 3.5)?

_____

b) Give the coordinates for …

Old Lighthouse. _____

Lookout Hill. _____

Clear Spring. _____

Feral Cat Island

c) Name the landmark …

1 km east of the Fort. _____

2 km south of Round Lake. _____

BONUS ▶ 1 km north and 1.5 km west of the Treasure. _____

d) Fill in the blanks.

From Round Lake, the Old Lighthouse is __5__ km ____east____.

From the Fort, walk _____ km _____ to the Treasure.

From the Treasure, the Bear Cave is _____ km _____.

To walk from the Bear Cave to Lookout Hill, walk _____ km _____ and _____ km south.

From the Old Lighthouse, walk _____ km _____ and _____ km _____ to the Clear Spring.

From the Fort, walk _____ to the Bear Cave.

From Lookout Hill to the Treasure, walk _____.

e) Write your own question that asks for directions and uses the map. Ask your partner to answer it.

_____

_____

G5-5 Graphing Sequences

REMINDER ▶ You can make sequences using multiplication and division, too.

Example: Start at 2. Multiply by 3 each time. Sequence: 2, 6, 18, 54, …

1. Extend both sequences.

a)

Divide by 3	Add 3
54	0
18	3
6	6
2	9

b)

Add 5	Multiply by 2
2	4

c)

Subtract 3	Divide by 2
25	88

BONUS ▶ Extend the table another row.

2. Extend both sequences. Then write a list of ordered pairs for the table.

a)

Add 5	Add 6	Ordered Pair
0	1	(0, 1)
5	7	(5, 7)
10	13	(,)

b)

Add 2	Add 2	Ordered Pair
3	1	

c)

Subtract 5	Subtract 6	Ordered Pair
100	66	

d)

Add 3	Multiply by 2	Ordered Pair
0	1	

e)

Add 1	Add 5	Ordered Pair
1	8	

f)

Add 4	Subtract 4	Ordered Pair
3	1	

3. a) Extend the sequences. Write a list of ordered pairs for the table.

i)

Add 3	Add 2	Ordered Pair
1	2	(1, 2)
4	4	(,)
7	6	(,)
10	8	(,)

ii)

Subtract 2	Add 2	Ordered Pair
9	0	

b) Plot the ordered pairs from part a) and connect the points.

i)
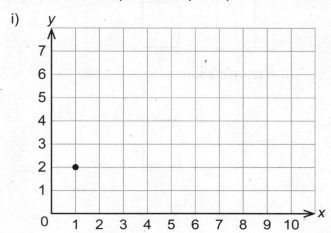

ii)
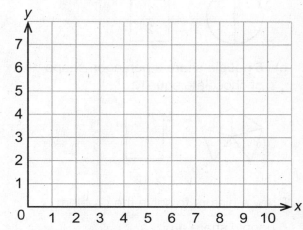

c) What do you notice about the points in each graph?

4. a) Extend the sequences. Write a list of ordered pairs for the table.

i)

Subtract 3	Subtract 2	Ordered Pair
12	7	(12, 7)

ii)

Multiply by 2	Add 2	Ordered Pair
0	1	

b) Plot the ordered pairs from part a) and connect the points.

i)
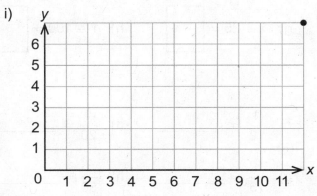

ii)
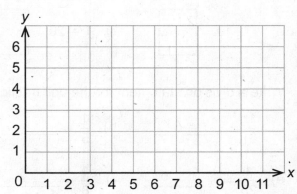

c) Which graph is different from the others on this page? How is one of the rules making it different?

NF5-20 Fractions and Division

Sarah wants to share a pie equally among four friends.

Each friend gets a quarter (or $\frac{1}{4}$) of the pie.

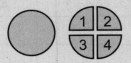

1. Shade how much one person gets. Write the fraction in the box.

   a) 2 people share a pancake equally.

   b) 3 people share a gold bar equally.

   c) 5 people share a pentagon shape of chocolate equally.

   d) 8 people share a pizza equally.

2. Draw a picture to solve the problem.

   6 people share a pizza.

   How much pizza does each person get? _____

You can use the division sign ($\div$) for equal sharing, even when the answer is a fraction.

Example: When 3 people share a pancake equally, each person gets $\frac{1}{3}$ of a pancake. So $1 \div 3 = \frac{1}{3}$.

---

3. Write a unit fraction for the division statement.

   a) $1 \div 3 = \boxed{\frac{1}{3}}$ b) $1 \div 4 =$ c) $1 \div 2 =$ d) $1 \div 8 =$

   e) $1 \div 10 =$ f) $1 \div 20 =$ g) $1 \div 25 =$ h) $1 \div 100 =$

4. Write a division statement for the unit fraction.

   a) $\frac{1}{6} = \underline{\ 1 \div 6\ }$ b) $\frac{1}{5} = \underline{\hspace{2cm}}$ c) $\frac{1}{12} = \underline{\hspace{2cm}}$ d) $\frac{1}{20} = \underline{\hspace{2cm}}$

   e) $\frac{1}{25} = \underline{\hspace{2cm}}$ f) $\frac{1}{8} = \underline{\hspace{2cm}}$ g) $\frac{1}{15} = \underline{\hspace{2cm}}$ h) $\frac{1}{50} = \underline{\hspace{2cm}}$

Problem: How can 4 people share 3 pies equally?

Solution: Share each pie equally.

One person takes the shaded pieces.

There are 4 people, so cut each pie into 4 pieces.

5. Determine the number of pieces and the number of whole pies.

   a) 3 people share 2 pies.

   Number of pieces in each pie: __3__

   Number of whole pies: __2__

   b) 2 people share 5 pies.

   Number of pieces in each pie: _____

   Number of whole pies: _____

   c) 3 people share 4 pies.

   Number of pieces in each pie: _____

   Number of whole pies: _____

   d) 5 people share 3 pies.

   Number of pieces in each pie: _____

   Number of whole pies: _____

6. Color one person's share of the pancakes. How much does each person get?

   a) 2 people share 3 pancakes.

   Each person gets $\dfrac{3}{2}$.

   b) 3 people share 2 pancakes.

   Each person gets ☐ .

   c) 4 people share 2 pancakes.

   Each person gets ☐ .

   d) 5 people share 3 pancakes.

   Each person gets ☐ .

7. Draw a picture to solve the problem.

   3 people share 5 pizzas.

   How much pizza does each person get? _____

Four friends share 3 pies equally. Each friend gets 3 quarters of a pie, so $3 \div 4 = \dfrac{3}{4}$.

 $\dfrac{3}{4}$ for the first friend $\dfrac{3}{4}$ for the third friend

3 pies for 4 friends $\dfrac{3}{4}$ for the second friend $\dfrac{3}{4}$ for the fourth friend

8. Write a fraction for the division statement.

a) $2 \div 7 = \boxed{\dfrac{2}{7}}$ b) $4 \div 5 = \boxed{}$ c) $3 \div 8 = \boxed{}$ d) $5 \div 9 = \boxed{}$

e) $5 \div 11 = \boxed{}$ f) $9 \div 10 = \boxed{}$ g) $10 \div 11 = \boxed{}$ h) $15 \div 22 = \boxed{}$

i) $23 \div 8 = \boxed{}$ j) $32 \div 25 = \boxed{}$ k) $43 \div 20 = \boxed{}$ l) $173 \div 100 = \boxed{}$

m) $19 \div 12 = \boxed{}$ n) $88 \div 50 = \boxed{}$ o) $56 \div 25 = \boxed{}$ p) $67 \div 10 = \boxed{}$

9. Write your answers to Question 8 parts i) to p) as mixed numbers.

i) $23 \div 8 = \boxed{2\dfrac{7}{8}}$ j) $32 \div 25 = \boxed{}$ k) $43 \div 20 = \boxed{}$ l) $173 \div 100 = \boxed{}$

m) $19 \div 12 = \boxed{}$ n) $88 \div 50 = \boxed{}$ o) $56 \div 25 = \boxed{}$ p) $67 \div 10 = \boxed{}$

10. Write a division statement for the fraction. Then find the answer.

a) $\dfrac{6}{3} = \underline{\quad 6 \div 3 \quad} = \underline{\quad 2 \quad}$ b) $\dfrac{12}{4} = \underline{\qquad\qquad} = \underline{\quad}$

c) $\dfrac{15}{3} = \underline{\qquad\qquad} = \underline{\quad}$ d) $\dfrac{24}{6} = \underline{\qquad\qquad} = \underline{\quad}$

e) $\dfrac{24}{4} = \underline{\qquad\qquad} = \underline{\quad}$ f) $\dfrac{25}{5} = \underline{\qquad\qquad} = \underline{\quad}$

g) $\dfrac{36}{9} = \underline{\qquad\qquad} = \underline{\quad}$ h) $\dfrac{56}{8} = \underline{\qquad\qquad} = \underline{\quad}$

11. Three friends want to share a 20-pound bag of rice equally by weight. How many pounds of rice should each friend get? Write your answer as a mixed number.

1. Extend the horizontal lines to find out what fraction of the whole rectangle is shaded.
    Then write the fraction.

    a) 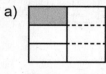   b)    c)   d)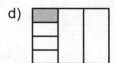

    $\dfrac{1}{6}$

    _____   _____   _____   _____

Here is $\dfrac{1}{2}$ of a rectangle. | Here is $\dfrac{1}{3}$ of $\dfrac{1}{2}$ of the rectangle. | How much is $\dfrac{1}{3}$ of $\dfrac{1}{2}$?

Extend the lines to find out.

$\dfrac{1}{3}$ of $\dfrac{1}{2} = \dfrac{1}{6}$

---

2. Extend the horizontal lines in the picture. Then complete a fraction statement for the
    picture using the word "of."

    a)   b)   c)   d)

    $\dfrac{1}{3}$ of $\dfrac{1}{4} = \dfrac{1}{12}$ $\dfrac{1}{3}$ of $\dfrac{1}{5} = $ —— $\dfrac{1}{2}$ of $\dfrac{1}{3} = $ —— $\dfrac{1}{4}$ of $\dfrac{1}{3} = $ ——

    e)   f)   g)   h)

    $\dfrac{1}{5}$ of $\dfrac{1}{2} = $ —— $\dfrac{1}{4}$ of $\dfrac{1}{5} = $ —— $\dfrac{1}{5}$ of $\dfrac{1}{3} = $ —— $\dfrac{1}{4}$ of $\dfrac{1}{2} = $ ——

REMINDER ▶ The word "of" can mean "multiply."

Example: "$\dfrac{1}{2}$ of a group of 8" means $\dfrac{1}{2} \times 8 = 8 \div 2 = 4$.

---

3. Rewrite the fraction statements from Question 2 using the multiplication sign instead
    of the word "of."

    a) $\dfrac{1}{3} \times \dfrac{1}{4} = \dfrac{1}{12}$ b)   c)   d)

    e)   f)   g)   h)

4. Extend the horizontal lines. Then write a multiplication statement for the picture.

a)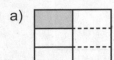

$$\frac{1}{3} \times \frac{1}{2} = \frac{1}{6}$$

b)

c)

d)

e)

f)

g)

h)

Look at Question 4 a). How do you get the denominator of the answer from the other two denominators?

$$\frac{1}{3} \times \frac{1}{2} = \frac{1}{6} \longleftarrow 3 \times 2$$

---

5. Multiply.

a) $\dfrac{1}{2} \times \dfrac{1}{3} =$

b) $\dfrac{1}{4} \times \dfrac{1}{5} =$

c) $\dfrac{1}{5} \times \dfrac{1}{6} =$

d) $\dfrac{1}{7} \times \dfrac{1}{3} =$

e) $\dfrac{1}{3} \times \dfrac{1}{5} =$

f) $\dfrac{1}{5} \times \dfrac{1}{5} =$

g) $\dfrac{1}{5} \times \dfrac{1}{4} =$

h) $\dfrac{1}{5} \times \dfrac{1}{2} =$

i) $\dfrac{1}{4} \times \dfrac{1}{7} =$

j) $\dfrac{1}{6} \times \dfrac{1}{2} =$

k) $\dfrac{1}{3} \times \dfrac{1}{3} =$

l) $\dfrac{1}{7} \times \dfrac{1}{5} =$

6. Circle two multiplication statements in Question 5 that have the same answer.
How could you have predicted this?

---

7. Mike is making $\dfrac{1}{2}$ of a recipe for raisin bread.

The recipe calls for $\dfrac{1}{3}$ of a cup of raisins.

What fraction of a cup of raisins does Mike need?

8. There is $\dfrac{1}{3}$ of a pizza left. Clara eats $\dfrac{1}{4}$ of it.

What fraction of the whole pizza did Clara eat?

NF5-22 Multiplying Fractions

Here is $\frac{2}{3}$ of a rectangle.

Here is $\frac{4}{5}$ of $\frac{2}{3}$ of the rectangle.

How much is $\frac{4}{5}$ of $\frac{2}{3}$?

Extend the lines to find out.

$\frac{4}{5}$ of $\frac{2}{3} = \frac{8}{15}$

1. Extend the horizontal lines in the picture. Then write a fraction statement for the picture using the word "of."

a)

$\frac{3}{4}$ of $\frac{2}{5} = \frac{6}{20}$

b)

—— of —— = ——

c)

—— of —— = ——

d)

—— of —— = ——

e)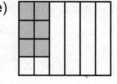

—— of —— = ——

f)

—— of —— = ——

g)

—— of —— = ——

h)

—— of —— = ——

$\frac{4}{5}$ of $\frac{2}{3} = \frac{8}{15}$ ⟵ —— 4×2

⟵ —— 5×3

2. Rewrite the fraction statements from Question 1 using the multiplication sign instead of the word "of."

a) $\frac{3}{4} \times \frac{2}{5} = \frac{6}{20}$

b)

c)

d)

e)

f)

g)

h)

3. Multiply.

a) $\frac{2}{3} \times \frac{4}{7} = \frac{8}{21}$

b) $\frac{1}{2} \times \frac{3}{5} =$

c) $\frac{3}{4} \times \frac{5}{7} =$

d) $\frac{2}{3} \times \frac{10}{11} =$

e) $\frac{3}{4} \times \frac{3}{5} =$

f) $\frac{2}{5} \times \frac{4}{7} =$

BONUS ▶ $\frac{1}{2} \times \frac{3}{5} \times \frac{3}{7} =$

4. Ron is making $\frac{1}{2}$ of a recipe for pasta salad. The recipe calls for $\frac{3}{4}$ of a cup of bow tie pasta. What fraction of a cup of bow tie pasta does Ron need?

5. There is $\frac{3}{8}$ of a pie left. Grace eats $\frac{3}{5}$ of it. What fraction of the whole pie did Grace eat?

6. Sharira spends $\frac{3}{5}$ of her free time playing outside. She spends $\frac{2}{3}$ of her outside time playing soccer. What fraction of her free time did she play soccer?

You can multiply *improper fractions* the same way you multiply *proper fractions*.

$$\frac{5}{2} =$$

$$\frac{3}{4} \times \frac{5}{2} =$$ $$= \frac{15}{8}$$ ← 5 groups of 3 are shaded.
← 2 groups of 4 are in each whole.

7. Multiply. Reduce your answer to lowest terms.

a) $\frac{2}{3} \times \frac{9}{4} = \frac{2 \times 9}{3 \times 4} = \frac{18}{12} = \frac{3}{2}$

b) $\frac{3}{4} \times \frac{12}{7} =$

c) $\frac{1}{2} \times \frac{8}{5} =$

d) $\frac{3}{2} \times \frac{6}{7} =$

e) $\frac{8}{3} \times \frac{7}{4} =$

f) $\frac{3}{5} \times \frac{15}{6} =$

g) $\frac{1}{6} \times \frac{12}{5} =$

BONUS ▶ $\frac{11}{5} \times \frac{10}{33} =$

8. Ben believes $\frac{2}{3}$ of $\frac{8}{5}$ is greater than 1. Is he right? Explain.

9. Farah made apple juice. She used $\frac{3}{5}$ of a bag of apples on Saturday. She used $\frac{1}{2}$ of the rest of the apples on Sunday. What fraction of the bag did Farah use on Sunday? Reduce your answer to lowest terms.

Number and Operations—Fractions 5-22

NF5-23 Multiplying Mixed Numbers

1. Write the mixed number as an improper fraction.

a) $3\frac{1}{2} = \frac{7}{2}$ ← $3 \times 2 + 1$

b) $4\frac{1}{3} =$

c) $2\frac{3}{5} =$

d) $1\frac{4}{7} =$

Megan multiplies $3 \times 2\frac{1}{4}$ in three steps.

Step 1

Change the mixed number to an improper fraction.

Example:

$3 \times 2\frac{1}{4} = 3 \times \frac{9}{4}$ ← $2 \times 4 + 1$

Step 2

Multiply the improper fraction by the whole number.

$3 \times \frac{9}{4} = \frac{27}{4}$ ← 9×3

Step 3

Change the improper fraction to a mixed number.

$\frac{27}{4} = 6\frac{3}{4}$ ← $27 \div 4 = 6\ R\ 3$

---

2. Change the mixed number to an improper fraction and multiply. Write your answer as a mixed number.

a) $2 \times 3\frac{2}{5} = 2 \times \frac{17}{5}$

$= \boxed{\dfrac{34}{5}}$ ← improper fraction

$= \boxed{6\dfrac{4}{5}}$ ← mixed number

b) $3 \times 4\frac{1}{2} =$

$= \boxed{}$ ← improper fraction

$= \boxed{}$ ← mixed number

c) $4 \times 1\frac{1}{3} =$

$= \boxed{}$ ← improper fraction

$= \boxed{}$ ← mixed number

d) $2\frac{3}{4} \times 6 =$

$= \boxed{}$ ← improper fraction

$= \boxed{}$ ← mixed number

e) $4\frac{2}{3} \times 2 =$

$= \boxed{}$ ← improper fraction

$= \boxed{}$ ← mixed number

f) $4 \times 1\frac{3}{5} =$

$= \boxed{}$ ← improper fraction

$= \boxed{}$ ← mixed number

g) $4 \times 2\frac{3}{10} =$

h) $3\frac{1}{5} \times 2 =$

i) $5\frac{1}{4} \times 3 =$

j) $2\frac{1}{6} \times 3 =$

Multiply two mixed numbers in three steps.

Step 1

Change the mixed numbers to improper fractions.

Example:

$$2\frac{3}{4} \times 1\frac{1}{2} = \frac{11}{4} \times \frac{3}{2}$$

$2 \times 4 + 3$

Step 2

Multiply the improper fractions.

$$\frac{11}{4} \times \frac{3}{2} = \frac{33}{8}$$

11×3

4×2

Step 3

Change the improper fraction to a mixed number.

$$\frac{33}{8} = 4\frac{1}{8}$$

$33 \div 8 = 4 \text{ R } 1$

3. Change the mixed numbers to improper fractions and multiply. Write your answer as a mixed number.

a) $1\frac{1}{3} \times 1\frac{3}{5} = \frac{4}{3} \times \frac{8}{5}$

$= \boxed{\dfrac{32}{15}}$ ← improper fraction

$= \boxed{2\dfrac{2}{15}}$ ← mixed number

b) $2\frac{3}{4} \times 3\frac{1}{2} =$

$= \boxed{}$ ← improper fraction

$= \boxed{}$ ← mixed number

c) $3\frac{1}{2} \times 2\frac{2}{3} =$

$= \boxed{}$ ← improper fraction

$= \boxed{}$ ← mixed number

d) $4\frac{2}{3} \times 2\frac{1}{5} =$

$= \boxed{}$ ← improper fraction

$= \boxed{}$ ← mixed number

e) $3\frac{1}{3} \times 2\frac{7}{10} =$ f) $1\frac{2}{3} \times 2\frac{1}{5} =$ g) $2\frac{1}{4} \times 1\frac{2}{5} =$ h) $2\frac{1}{3} \times 2\frac{2}{5} =$

4. Luis is making $\frac{3}{5}$ of a recipe for mushroom soup. The recipe calls for $3\frac{1}{2}$ cups of milk

a) How much milk does he need? Hint: Change $3\frac{1}{2}$ to an improper fraction.

b) Luis uses 2 cups of milk. Will his recipe work?

5. Nina is making $3\frac{1}{2}$ batches of cookies. The recipe for one batch calls for $1\frac{3}{4}$ cups of flour.

a) How many cups of flour does Nina need?

b) She has 6 cups of flour. If she uses all her flour, will the recipe work?

Number and Operations—Fractions 5-23

NF5-24 Multiplication and Fractions

1. Find the product. The fractions in the questions are all *smaller than 1*. Compare your answer to the whole number.

   a) $\frac{2}{3} \times 6 = \frac{12}{3} = 4$

   $\frac{2}{3} \times 6$ is ___*smaller*___ than 6.

   b) $\frac{3}{4} \times 8 = \frac{\quad}{4} =$

   $\frac{3}{4} \times 8$ is _____ than 8.

   c) $\frac{2}{5} \times 10 = \frac{\quad}{5} =$

   $\frac{2}{5} \times 10$ is _____ than 10.

   d) $\frac{5}{6} \times 12 = \frac{\quad}{6} =$

   $\frac{5}{6} \times 12$ is _____ than 12.

   e) $\frac{4}{8} \times 6 = \frac{\quad}{8} =$

   $\frac{4}{8} \times 6$ is _____ than 6.

   f) $\frac{3}{12} \times 8 = \frac{\quad}{12} =$

   $\frac{3}{12} \times 8$ is _____ than 8.

2. When you multiply a whole number by a fraction smaller than 1, do you think the answer will be less than the whole number or greater than the whole number?

   _____

To find $\frac{2}{3} \times 9$ or $\frac{2}{3}$ of 9 …

Step 1: Divide 9 into 3 equal parts.

Step 2: Select 2 of the parts.

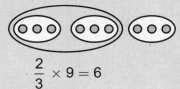

$\frac{2}{3} \times 9 = 6$

---

3. Draw a picture to find the product.

   a) $\frac{3}{4} \times 12$ or $\frac{3}{4}$ of 12

   b) $\frac{2}{3} \times 6$ or $\frac{2}{3}$ of 6

   c) $\frac{3}{4} \times 8$ or $\frac{3}{4}$ of 8

   d) $\frac{4}{5} \times 15$ or $\frac{4}{5}$ of 15

   e) $\frac{5}{6} \times 12$ or $\frac{5}{6}$ of 12

4. State the meaning of the product.

   a) $\frac{3}{5} \times 15$

   Divide __15__ into __5__ parts.

   Select __3__ of the parts.

   b) $\frac{2}{3} \times 12$

   Divide _____ into ____ parts.

   Select ____ of the parts.

   c) $\frac{2}{5} \times 10$

   Divide _____ into ____ parts.

   Select ____ of the parts.

5. Draw a picture to show why $\frac{4}{5}$ of 20 is less than 20.

6. Find the product. The fractions in the questions are all *greater than 1*. Compare your answer to the whole number.

a) $\frac{5}{4} \times 8 = \frac{40}{4} = 10$

$\frac{5}{4} \times 8$ is ___*larger*___ than 8.

b) $\frac{4}{3} \times 6 = \frac{}{3} =$

$\frac{4}{3} \times 6$ is _____ than 6.

c) $\frac{6}{5} \times 10 = \frac{}{5} =$

$\frac{6}{5} \times 10$ is _____ than 10.

d) $\frac{8}{6} \times 12 = \frac{}{6} =$

$\frac{8}{6} \times 12$ is _____ than 12.

e) $\frac{5}{2} \times 6 = \frac{}{2} =$

$\frac{5}{2} \times 6$ is _____ than 6.

f) $\frac{10}{9} \times 18 = \frac{}{9} =$

$\frac{10}{9} \times 18$ is _____ than 18.

7. When you multiply a whole number by a fraction greater than 1, do you think the answer will be less than the whole number or greater than the whole number?

_____

8. Rewrite the product in expanded form.

a) $1\frac{1}{2} \times 3 = \left(1 + \frac{1}{2}\right) \times 3$

$= 1 \times 3 + \frac{1}{2} \times 3$

b) $1\frac{3}{4} \times 5 = \left(\underline{} + \underline{}\right) \times \underline{}$

$= \underline{} + \underline{}$

c) $1\frac{5}{6} \times 2 = \left(\underline{} + \underline{}\right) \times \underline{}$

$= \underline{} + \underline{}$

9. Explain why $1\frac{1}{2} \times 3$ must be greater than 3. Hint: Use your work from Question 8.

10. Explain why $\frac{5}{4} \times 7$ must be greater than 7. Hint: Change $\frac{5}{4}$ to a mixed number.

11. Predict whether the product will be greater than or less than $\frac{1}{2}$. Then find the product.

a) $\frac{2}{3} \times \frac{1}{2}$ is _____ than $\frac{1}{2}$

$\frac{2}{3} \times \frac{1}{2} = \boxed{}$

b) $\frac{6}{5} \times \frac{1}{2}$ is _____ than $\frac{1}{2}$

$\frac{6}{5} \times \frac{1}{2} = \boxed{}$

12. Alex has 12 stickers. Maria has $1\frac{1}{4}$ times as many as Alex. Sofia has $\frac{5}{6}$ as many as Alex.

a) Without calculating, say who has the greatest number of stickers. Explain your answer.

b) To check your answer in part a), calculate the number of stickers for Maria and Sofia.

REMINDER ▶ To find a part of a whole, you can use a scale factor.

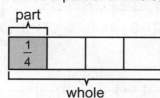

part = whole × $\frac{1}{4}$ Scale factor is $\frac{1}{4}$.

whole = part × 4   Scale factor is 4.

1. Using a ruler and the scale factor given, extend the drawing to show the whole.

a) $\frac{1}{2}$

Scale factor = 2

b) $\frac{1}{3}$

Scale factor = 3

c) $\frac{1}{4}$

Scale factor = 4

d) $\frac{1}{5}$

Scale factor = 5

2. Using a ruler and the scale factor given, shade to show the part.

a)

Scale factor = $\frac{1}{2}$

b)

Scale factor = $\frac{1}{3}$

3. Using a ruler, find what fraction of the box is shaded. What is the scale factor?

a)

$\frac{1}{3}$ is shaded

whole = part × ___3___

part = whole × $\boxed{\frac{1}{3}}$

b)

___ is shaded

whole = part × _____

part = whole × $\boxed{}$

c)

___ is shaded

whole = part × _____

part = whole × $\boxed{}$

d)

___ is shaded

whole = part × _____

part = whole × $\boxed{}$

4. For the bold number, circle the number that is $\frac{1}{2}$ and draw a box around the number that is double.

a)
```
 ├──┼──┼──┼──┼──┼──┼──┼──┼──┼──┼──┼──┤
 0  ①  2  3  4  5  6  7  8  9 10 11 12
```
(1 is circled, 4 is boxed)

b)
```
 ├──┼──┼──┼──┼──┼──┼──┼──┼──┼──┼──┼──┤
 0  1  2  3  4  5  6  7  8  9 10 11 12
```

c)
```
 ├──┼──┼──┼──┼──┼──┼──┼──┼──┼──┼──┼──┤
 0  1  2  3  4  5  6  7  8  9 10 11 12
```

BONUS ▶
```
 ├──┼──┼──┼──┼──┼──┼──┼──┼──┼──┼──┼──┤
 0  ½  1 1½  2 2½  3 3½  4 4½  5 5½  6
```

5. Multiply 6 by the scale factor. Circle the answer on the number line.

a) Scale factor: $\frac{1}{2}$
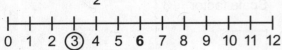
```
 ├──┼──┼──┼──┼──┼──┼──┼──┼──┼──┼──┼──┤
 0  1  2  ③  4  5  6  7  8  9 10 11 12
```

b) Scale factor: $\frac{1}{3}$
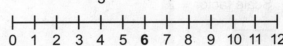
```
 ├──┼──┼──┼──┼──┼──┼──┼──┼──┼──┼──┼──┤
 0  1  2  3  4  5  6  7  8  9 10 11 12
```

c) Scale factor: 2
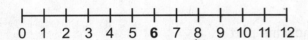
```
 ├──┼──┼──┼──┼──┼──┼──┼──┼──┼──┼──┼──┤
 0  1  2  3  4  5  6  7  8  9 10 11 12
```

d) Scale factor: $1\frac{1}{2}$ or $\frac{3}{2}$
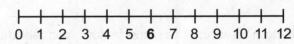
```
 ├──┼──┼──┼──┼──┼──┼──┼──┼──┼──┼──┼──┤
 0  1  2  3  4  5  6  7  8  9 10 11 12
```

e) Scale factor: $\frac{2}{3}$
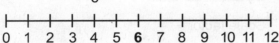
```
 ├──┼──┼──┼──┼──┼──┼──┼──┼──┼──┼──┼──┤
 0  1  2  3  4  5  6  7  8  9 10 11 12
```

f) Scale factor: $1\frac{1}{3}$ or $\frac{4}{3}$
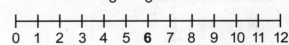
```
 ├──┼──┼──┼──┼──┼──┼──┼──┼──┼──┼──┼──┤
 0  1  2  3  4  5  6  7  8  9 10 11 12
```

REMINDER ▶ Multiplying by a scale factor smaller than 1 makes the number smaller.

Multiplying by a scale factor larger than 1 makes the number larger.

---

6. Multiply the number by the scale factor.

a) 5 by the scale factor $\frac{1}{2}$

b) 7 by the scale factor $\frac{1}{3}$

$$5 \times \frac{1}{2} = \frac{5}{2} = 2\frac{1}{2}$$

c) $\frac{1}{2}$ by the scale factor 3

d) 3 by the scale factor 2

BONUS ▶ On a map, 1 cm represents 500 m.

a) If a lake is 6 cm long on the map, what is the actual size of the lake?

b) The distance between the forest and the ocean in real life is 1,000 m. What is the distance on the map?

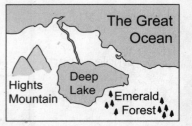

NF5-26 Word Problems with Fractions and Multiplication

1. A foot is 12 inches. A yard is 3 feet. What fraction of a yard is 8 inches?

   Find what fraction of a foot 8 inches is: *8 inches is $\frac{8}{12}$ of a foot.*

   Find what fraction of a yard a foot is: *1 foot is $\frac{1}{3}$ of a yard.*

   So 8 inches $= \frac{8}{12} \times \frac{1}{3}$ of a yard $=$ _____ of a yard.

---

2. A year has _____ months.

   A decade has _____ years.

   What fraction of a decade is 8 months? _____

3. A century has 100 years. A year has 52 weeks.

   What fraction of a century is 13 weeks? _____

---

4. A day has 24 hours. An hour has 60 minutes.
   What fraction of a day is 45 minutes?

   BONUS ▶ What fraction of a day is 40 seconds?

5. How many months old is a $1\frac{2}{3}$-year-old child?

---

6. a) Anwar has $\frac{7}{5}$ cups of flour. He uses $\frac{3}{4}$ of it to bake a cake. How much flour did he use?

   b) Did Anwar use more or less than 1 cup of flour? How do you know?

7. Kim has $\frac{5}{3}$ cups of paint. She uses $\frac{3}{4}$ of it to paint a shelf.

   a) How much paint did she use?

   b) Did Kim use more or less than 1 cup of paint? How do you know?

---

8. Leo has $\frac{11}{5}$ teaspoons of salt. He uses $\frac{1}{3}$ of it to cook a pot of soup. He then eats $\frac{2}{9}$ of the soup. How much salt did Leo eat?

BONUS ▶ The peel of a banana weighs $\frac{1}{8}$ of the total weight of a banana. You buy 6 kg of bananas at $0.50 per kg. How much do you pay for the peel?

---

1. One half of the whole rectangle is shaded. What fraction of the whole rectangle is striped? Hint: How many striped pieces fit into the whole rectangle?

a)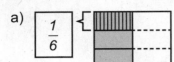

b)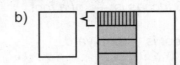

c)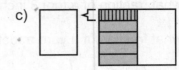

2. What is the area of the striped part? Hint: Extend the lines to divide the rectangle into smaller parts.

a) $\frac{1}{12}$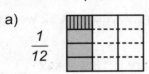

b) ___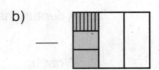

c) ___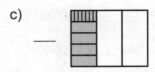

d) ___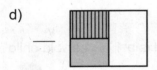

e) ___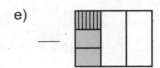

f) ___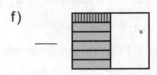

g) ___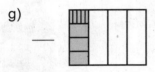

h) ___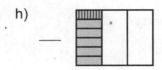

i) ___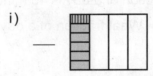

j) ___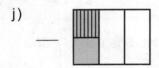

k) ___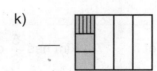

l) ___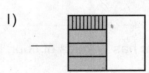

The model in Question 2.a) shows that $\frac{1}{3} \div 4 = \frac{1}{12}$

because $\frac{1}{3}$ of the whole rectangle is divided into 4 smaller parts.

3. Write a division statement for each model in Question 2.

a) $\frac{1}{3} \div 4 = \frac{1}{12}$

b)

c)

d)

e)

f)

g)

h)

i)

j)

k)

l)

Five people share $\frac{2}{3}$ of a cake. What fraction of the cake does each person get?

Divide $\frac{2}{3}$ into 5 equal groups. How much is in each group?

 ÷ 5 = → 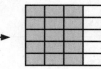 so $\frac{2}{3} \div 5 = \frac{2}{15}$

4. Use the model to divide each fraction.

a)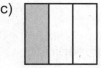

$\frac{3}{4} \div 5 = \frac{3}{20}$

b)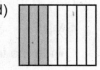

$\frac{1}{2} \div 3 =$

c)

$\frac{1}{3} \div 4 =$

d)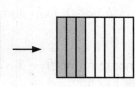

$\frac{3}{8} \div 5 =$

5. Finish the model to divide.

a)

$\frac{1}{2} \div 3 = \frac{1}{6}$

b)

$\frac{2}{3} \div 5 =$

c)

$\frac{1}{3} \div 4 =$

d)

$\frac{3}{5} \div 4 =$

6. Check your answers to Question 5 using multiplication. Example: Does $3 \times \frac{1}{6} = \frac{1}{2}$?

$$\frac{2}{5} \div 3 = \frac{2}{5 \times 3} = \frac{2}{15}$$

7. Divide. Check your answers using a model.

a) $\frac{3}{5} \div 2 =$ b) $\frac{1}{7} \div 4 =$ c) $\frac{4}{5} \div 3 =$ d) $\frac{2}{3} \div 3 =$

8. Five people share $\frac{5}{8}$ of a cake equally. What fraction of the cake does each person eat?

9. Four people share $\frac{1}{2}$ lb of almonds equally. How much does each person get?

You can divide improper fractions by whole numbers the same way you divide proper fractions.

$$\frac{5}{2} =$$

$$\frac{5}{2} \div 4 = \qquad = \frac{5}{8} \longleftarrow \qquad 2 \times 4$$

10. Divide.

a) $\dfrac{5}{4} \div 2 = \dfrac{5}{4 \times 2} = \dfrac{5}{8}$

b) $\dfrac{5}{2} \div 3 = \dfrac{\qquad}{\qquad} = \dfrac{\ }{\ }$

c) $\dfrac{7}{4} \div 5 = \dfrac{\qquad}{\qquad} = \dfrac{\ }{\ }$

d) $\dfrac{8}{3} \div 6 = \dfrac{\qquad}{\qquad} = \dfrac{\ }{\ }$

Divide mixed numbers by whole numbers in three steps.

Step 1

Change the mixed number to an improper fraction.

Example: $\overset{5 \times 4 + 3}{\nearrow}$

$5\dfrac{3}{4} \div 2 = \dfrac{23}{4} \div 2$

Step 2

Divide the improper fraction.

$\dfrac{23}{4} \div 2 = \dfrac{23}{4 \times 2} = \dfrac{23}{8}$

Step 3

Change the improper fraction to a mixed number.

$\overset{23 \div 8 = 2 \ R \ 7}{\nearrow}$

$\dfrac{23}{8} = 2\dfrac{7}{8}$

11. Divide by changing the mixed number to an improper fraction.

a) $2\dfrac{3}{4} \div 5 = \dfrac{11}{4} \div 5 = \dfrac{11}{4 \times 5} = \dfrac{11}{20}$

b) $2\dfrac{1}{2} \div 3 = \dfrac{\ }{\ } \div 3 = \dfrac{\qquad}{\qquad} = \dfrac{\ }{\ }$

c) $1\dfrac{3}{4} \div 5 = \dfrac{\ }{\ } \div 5 = \dfrac{\qquad}{\qquad} = \dfrac{\ }{\ }$

d) $2\dfrac{2}{3} \div 6 = \dfrac{\ }{\ } \div 6 = \dfrac{\qquad}{\qquad} = \dfrac{\ }{\ }$

e) $3\dfrac{2}{5} \div 4 = \dfrac{\ }{\ } \div 4 = \dfrac{\qquad}{\qquad} = \dfrac{\ }{\ }$

f) $1\dfrac{4}{7} \div 2 = \dfrac{\ }{\ } \div 2 = \dfrac{\qquad}{\qquad} = \dfrac{\ }{\ }$

12. Three people share $1\dfrac{1}{2}$ pizzas equally. What fraction of the pizzas does each person eat?

BONUS ▶ Two people can paint a house in 5 days.

a) How long does it take 5 people to paint that house?

b) How long does it take 4 people to paint that house?

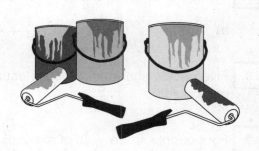

NF5-28 Dividing Whole Numbers by Unit Fractions

There are 4 quarters in one whole.

So $1 \div \dfrac{1}{4} = 4$.

1			
$\frac{1}{4}$	$\frac{1}{4}$	$\frac{1}{4}$	$\frac{1}{4}$

1. How many fit into 1?

a)

1	
$\frac{1}{2}$	$\frac{1}{2}$

The number of $\dfrac{1}{2}$s in a whole is ___2___.

b)

1				
$\frac{1}{5}$	$\frac{1}{5}$	$\frac{1}{5}$	$\frac{1}{5}$	$\frac{1}{5}$

The number of $\dfrac{1}{5}$s in a whole is _____.

c)

1		
$\frac{1}{3}$	$\frac{1}{3}$	$\frac{1}{3}$

The number of $\dfrac{1}{3}$s in a whole is _____.

d)

1					
$\frac{1}{6}$	$\frac{1}{6}$	$\frac{1}{6}$	$\frac{1}{6}$	$\frac{1}{6}$	$\frac{1}{6}$

The number of $\dfrac{1}{6}$s in a whole is _____.

2. Complete the division statement using your answers from Question 1.

a) $1 \div \dfrac{1}{2} = $ ___2___

b) $1 \div \dfrac{1}{5} = $ _____

c) $1 \div \dfrac{1}{3} = $ _____

d) $1 \div \dfrac{1}{6} = $ _____

There are 4 quarters in each whole, so there are 8 quarters in two wholes.

So $2 \div \dfrac{1}{4} = 8$.

1				1			
$\frac{1}{4}$	$\frac{1}{4}$	$\frac{1}{4}$	$\frac{1}{4}$	$\frac{1}{4}$	$\frac{1}{4}$	$\frac{1}{4}$	$\frac{1}{4}$

3. How many fit into 2?

a)

1		1	
$\frac{1}{2}$	$\frac{1}{2}$	$\frac{1}{2}$	$\frac{1}{2}$

$2 \div \dfrac{1}{2} = $ _____

b)

1			1		
$\frac{1}{3}$	$\frac{1}{3}$	$\frac{1}{3}$	$\frac{1}{3}$	$\frac{1}{3}$	$\frac{1}{3}$

$2 \div \dfrac{1}{3} = $ _____

c) $2 \div \dfrac{1}{5} = $ _____

d) $2 \div \dfrac{1}{10} = $ _____

4. Write a division statement for the picture.

a)

1		1		1	
$\frac{1}{2}$	$\frac{1}{2}$	$\frac{1}{2}$	$\frac{1}{2}$	$\frac{1}{2}$	$\frac{1}{2}$

$$3 \div \frac{1}{2} = 6$$

b)

1				1			
$\frac{1}{4}$	$\frac{1}{4}$	$\frac{1}{4}$	$\frac{1}{4}$	$\frac{1}{4}$	$\frac{1}{4}$	$\frac{1}{4}$	$\frac{1}{4}$

_____

c)

1					1					1				
$\frac{1}{5}$	$\frac{1}{5}$	$\frac{1}{5}$	$\frac{1}{5}$	$\frac{1}{5}$	$\frac{1}{5}$	$\frac{1}{5}$	$\frac{1}{5}$	$\frac{1}{5}$	$\frac{1}{5}$	$\frac{1}{5}$	$\frac{1}{5}$	$\frac{1}{5}$	$\frac{1}{5}$	$\frac{1}{5}$

_____

5. Use the number line to decide how many steps of size $\frac{1}{3}$ fit into 4.

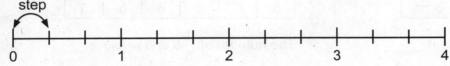

step

0   1   2   3   4

Write the division statement. _____

6. Draw a number line to divide 4 by $\frac{1}{2}$. Then write the division statement.

How many $\frac{1}{3}$s fit into 5? Three $\frac{1}{3}$s fit into 1, so 5 times as many fit into 5. $5 \div \frac{1}{3} = 5 \times 3 = 15$

7. Divide.

a) $2 \div \frac{1}{3} = \underline{\ 2\ } \times \underline{\ 3\ } = \underline{\ 6\ }$

b) $4 \div \frac{1}{3} = \underline{\quad} \times \underline{\quad} = \underline{\quad}$

c) $3 \div \frac{1}{2} = \underline{\quad} \times \underline{\quad} = \underline{\quad}$

d) $5 \div \frac{1}{2} = \underline{\quad} \times \underline{\quad} = \underline{\quad}$

e) $6 \div \frac{1}{4} = \underline{\quad} \times \underline{\quad} = \underline{\quad}$

f) $4 \div \frac{1}{6} = \underline{\quad} \times \underline{\quad} = \underline{\quad}$

8. a) Ethan has a scoop that measures a $\frac{1}{2}$ cup. He needs 3 cups of flour. How many scoopfuls of flour does he need?

b) Alice has 2 chocolate bars. She cuts each chocolate bar into quarters. How many pieces does she have?

NF5-29 Word Problems with Fractions and Division

1. Greg jogs $\frac{1}{2}$ km in 3 minutes. How far does he jog in 1 minute?

   $\frac{1}{2} \div 3 =$ _____ km

2. a) A cake recipe uses $\frac{3}{4}$ cups of sugar.
      The cake is divided into 6 pieces.
      How much sugar is in each piece?

   b) Another cake recipe uses $1\frac{1}{3}$ cups of sugar.
      The cake is divided into 8 pieces.
      How much sugar is in each piece?

3. Rosa invites friends for dinner. She has $\frac{6}{8}$ kg of dry spaghetti. Each person needs $\frac{3}{16}$ kg. How many people can she feed?

4. A rope is $4\frac{4}{5}$ m long. It is divided equally into 12 pieces. How long is each piece?

5. If 5 egg cartons weigh $3\frac{1}{2}$ pounds, find the weight of one carton.

6. a) What fraction of a year is a month?
   b) What fraction of a decade is a year?
   c) What fraction of a decade is a month?

7. A bookshelf is 26 inches wide. How many $\frac{1}{2}$-inch wide books can fit on the bookshelf?

BONUS ▶ A string is $3\frac{1}{2}$ m long. It is divided into 5 equal parts. How long is each piece?

$3\frac{1}{2} \div 5 = \frac{7}{2} \div 5 =$ _____ m = _____ cm

1. Craig has $48 dollars. He spends $\frac{1}{4}$ of his money on a book and $\frac{3}{8}$ on a pair of gloves.

   a) How many dollars did Craig spend on the gloves?

   b) How much money does he have left?

2. Ben and Kevin mow the grass in a backyard. Ben mows $\frac{3}{5}$ of the yard and Kevin mows the rest.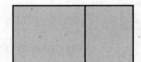

   a) What fraction of the yard did Kevin mow?

   b) The yard is 30 m long. How long is each person's section?

3. Sally stretches for $\frac{1}{2}$ an hour, walks for $\frac{2}{3}$ of an hour, and jogs for $\frac{2}{5}$ of an hour.

   Use two methods to calculate how many minutes she exercised.

   a) Add the fractions and then convert to minutes.

   b) Convert the fractions to minutes and then add.

   c) Did you get the same answers from parts a) and b)?

4. Ivan talks on the phone for $\frac{1}{5}$ of an hour. Jake talks on the phone for $\frac{1}{100}$ of a day.

   Who talked longer on the phone?

5. Sofia does one jumping jack in 2 seconds. How many jumping jacks can she do in

   $\frac{7}{10}$ of a minute? Circle the correct formula.

   $\frac{7}{10} \times 2$ $\frac{7}{10} \div 2$ $2 \div \frac{7}{10}$ $42 \div 2$ $2 \div 42$

   Explain your choice. _____

   _____

Use the information in this box to answer Questions 6 to 9.

Jade spends $\frac{3}{4}$ of an hour doing math homework, $\frac{2}{5}$ of an hour doing geography, and $\frac{5}{6}$ of an hour doing science.

---

6. Compare fractions to answer the questions.

   a) Did Jade spend more or less than half an hour doing geography?

   b) Did Jade spend more time doing math or science?

7. Decide whether the problem requires addition, subtraction, multiplication, or division. Then solve the problem.

a) Jade spends $\frac{1}{3}$ of her time doing math homework working on word problems. What fraction of an hour did she spend on word problems?

b) How many hours did Jade spend on homework altogether?

c) How many more hours did Jade spend on science than on geography?

d) Jade divides her time on geography evenly between reading the textbook and answering questions. What fraction of an hour did she spend reading the textbook?

8. Solve the problems.

a) Jade has 2 hours to do her homework before she has to leave to meet a friend. How many minutes before she has to leave did Jade finish her homework?

b) Jade spends $\frac{1}{3}$ of her time on science reading the textbook. She spends $\frac{1}{5}$ of her remaining time on science doing calculations. How many hours did she spend on calculations?

9. a) Jade's sister spends $\frac{1}{3}$ the amount of time Jade does on math, but twice the amount of time on geography and $1\frac{1}{2}$ times as much time on science.

Who spent more time doing homework, Jade or her sister?

b) Jade's brother spends the same total amount of time on homework as Jade's sister. He spends the same amount of time on all three subjects. How much time did Jade's brother spend on each subject? Write your answer as a fraction of an hour.

c) Jade's sister spends $\frac{1}{5}$ of her time on science reading the textbook.

Who spent more time reading her science textbook, Jade or her sister?

10. Make up a word problem that requires finding:

a) $\frac{4}{5} + \frac{1}{3}$ b) $\frac{4}{5} - \frac{1}{3}$ c) $\frac{4}{5} \times \frac{1}{3}$ d) $\frac{4}{5} \div \frac{1}{3}$

1. Name the shaded fraction.

   a)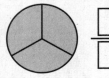

   b)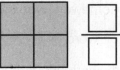

   c)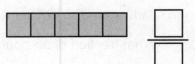

2. A fraction is equivalent to 1 if its numerator and denominator are _____.

3. Circle the fraction equivalent to 1. Then write if the fractions are "more" or "less" than 1.

   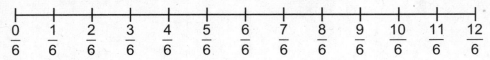

   a) $\frac{5}{6}$ is _____ than 1.

   b) $\frac{11}{6}$ is _____ than 1.

A fraction is less than 1 if the numerator is less than the denominator. Example: $\frac{2}{3}$ is less than 1.

A fraction is greater than 1 if the numerator is greater than the denominator. Example: $\frac{5}{4}$ is greater than 1.

---

4. Is the fraction less than or greater than 1? Write > (greater than) or < (less than).

   a) $\frac{7}{8}$ ☐ 1

   b) $\frac{4}{7}$ ☐ 1

   c) $\frac{14}{13}$ ☐ 1

   d) $\frac{23}{24}$ ☐ 1

5. Circle the fraction equivalent to $\frac{1}{2}$ on the number line.

   a)
   | 0 | $\frac{1}{4}$ | $\frac{2}{4}$ | $\frac{3}{4}$ | 1 |

   b)
   | 0 | $\frac{1}{6}$ | $\frac{2}{6}$ | $\frac{3}{6}$ | $\frac{4}{6}$ | $\frac{5}{6}$ | 1 |

A fraction is equal to half if double the numerator is equal to the denominator.

Example: $\frac{5}{10}$ is equal to $\frac{1}{2}$ because the double of 5 is equal to 10.

---

6. Fill in the blank.

   a) $\frac{1}{2} = \frac{7}{14}$

   b) $\frac{1}{2} = \frac{30}{}$

   c) $\frac{1}{2} = \frac{}{800}$

   d) $\frac{1}{2} = \frac{43}{}$

A fraction is less than half if double the numerator is less than the denominator.

Example: $\dfrac{4}{9}$ is less than $\dfrac{1}{2}$ because the double of 4 is less than 9.

A fraction is more than half if double the numerator is more than the denominator.

Example: $\dfrac{3}{5}$ is more than $\dfrac{1}{2}$ because the double of 3 is more than 5.

7. Write $>$ (greater than), $<$ (less than), or $=$ (equal to).

a) $\dfrac{3}{8}$ ☐ $\dfrac{1}{2}$ b) $\dfrac{4}{7}$ ☐ $\dfrac{1}{2}$ c) $\dfrac{13}{25}$ ☐ $\dfrac{1}{2}$ d) $\dfrac{1}{2}$ ☐ $\dfrac{23}{50}$

e) $\dfrac{18}{36}$ ☐ $\dfrac{1}{2}$ f) $\dfrac{9}{17}$ ☐ $\dfrac{1}{2}$ g) $\dfrac{1}{2}$ ☐ $\dfrac{14}{29}$ h) $\dfrac{37}{72}$ ☐ $\dfrac{1}{2}$

8. Write "greater" or "less."

a) $\dfrac{1}{4}$ is _____ than $\dfrac{1}{2}$ and $\dfrac{1}{2}$ is _____ than $\dfrac{4}{6}$, so $\dfrac{1}{4}$ is _____ than $\dfrac{4}{6}$.

b) $\dfrac{53}{100}$ is _____ than $\dfrac{1}{2}$ and $\dfrac{1}{2}$ is _____ than $\dfrac{3}{7}$, so $\dfrac{53}{100}$ is _____ than $\dfrac{3}{7}$.

9. Karen eats $\dfrac{3}{8}$ of a pizza. Is that more or less than half the pizza? _____

10. In a Grade 5 class, $\dfrac{4}{9}$ of the students are girls. Are there more girls or boys in the class? Explain.

11. On a baseball team, $\dfrac{6}{11}$ of the players are girls.

Are there more girls or boys on the team?

12. Ron eats $\dfrac{3}{5}$ of a pizza, and Karen eats $\dfrac{1}{3}$ of the pizza.

Who ate more pizza? Explain how you know.

13. Maria thinks that $\dfrac{4}{3}$ is less than $\dfrac{99}{100}$ because the numbers are smaller. Is she right?

Explain how you know.

14. Is $\dfrac{3}{10}$ of $\dfrac{4}{3}$ less than or greater than $\dfrac{1}{2}$? Explain.

NF5-32 Estimating with Fractions

1. Label the fraction on the number line with its letter.

A. halfway between $\frac{1}{4}$ and $\frac{3}{4}$

B. an equal distance from 1 and 2

C. halfway between 0 and $\frac{6}{4}$

D. an equal distance from $\frac{3}{4}$ and $\frac{5}{4}$

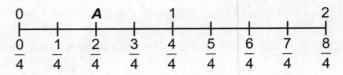

2. Draw an arrow to show whether each bold fraction is closest to 0, $\frac{1}{2}$, or 1.

a)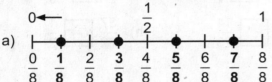

b)

c)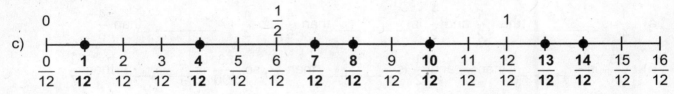

3. Estimate by rounding one of the fractions to 0, $\frac{1}{2}$, or 1.
(Hint: Use your answers in Question 2.)

a) $\dfrac{1}{12} + \dfrac{2}{3} \approx 0 + \dfrac{2}{3} = \dfrac{2}{3}$

b) $\dfrac{7}{6} \times \dfrac{13}{17} \approx$

c) $\dfrac{22}{75} - \dfrac{1}{12} \approx$

d) $\dfrac{7}{12} \div \dfrac{1}{2} \approx$

4. Is the fraction smaller than $\frac{1}{2}$ or larger than $\frac{1}{2}$?

(Hint: Double the numerator and compare it to the denominator.)

	Fraction	Double the Numerator	Smaller than $\frac{1}{2}$	Larger than $\frac{1}{2}$
a)	$\dfrac{5}{7}$	___5___ × ___2___ = ___10___ is _larger_ than ___7___		✓
b)	$\dfrac{4}{9}$	____ × ____ = ____ is _____ than ____		
c)	$\dfrac{11}{20}$	____ × ____ = ____ is _____ than ____		
d)	$\dfrac{14}{31}$	____ × ____ = ____ is _____ than ____		

$\frac{3}{7} + \frac{4}{9}$ is smaller than 1 because both fractions $\frac{3}{7}$ and $\frac{4}{9}$ are smaller than $\frac{1}{2}$.

$\frac{2}{3} + \frac{4}{7}$ is larger than 1 because both fractions $\frac{2}{3}$ and $\frac{4}{7}$ are larger than $\frac{1}{2}$.

5. Write "more" or "less."

a) $\frac{4}{6} + \frac{3}{5}$ is _____ than 1.

b) $\frac{1}{4} + \frac{2}{5}$ is _____ than 1.

c) $\frac{3}{7} + \frac{4}{10}$ is _____ than 1.

d) $\frac{3}{4} + \frac{3}{5}$ is _____ than 1.

e) $\frac{5}{9} + \frac{4}{7}$ is _____ than 1.

f) $\frac{2}{5} + \frac{3}{10}$ is _____ than 1.

6. Draw an arrow to show which whole number each bold fraction or mixed number is closer to.

a)
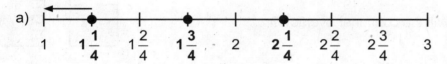

$1 \quad 1\frac{1}{4} \quad 1\frac{2}{4} \quad 1\frac{3}{4} \quad 2 \quad 2\frac{1}{4} \quad 2\frac{2}{4} \quad 2\frac{3}{4} \quad 3$

b)

$0 \quad \frac{1}{3} \quad \frac{2}{3} \quad 1 \quad 1\frac{1}{3} \quad 1\frac{2}{3} \quad 2$

c)

$4 \quad 4\frac{1}{5} \quad 4\frac{2}{5} \quad 4\frac{3}{5} \quad 4\frac{4}{5} \quad 5 \quad 5\frac{1}{5} \quad 5\frac{2}{5} \quad 5\frac{3}{5} \quad 5\frac{4}{5} \quad 6$

7. Round the mixed numbers to the nearest whole number.

	Mixed Number	Fraction Part	Double the Numerator	Smaller than $\frac{1}{2}$	Larger than $\frac{1}{2}$	Nearest Whole Number
a)	$4\frac{7}{10}$	$\frac{7}{10}$	_7_ × _2_ = _14_ is _larger_ than _10_		✓	5
b)	$6\frac{4}{9}$	$\frac{4}{9}$	_4_ × _2_ = _8_ is _smaller_ than _9_	✓		6
c)	$4\frac{3}{7}$		___ × ___ = ___ is _____ than ___			
d)	$2\frac{9}{15}$		___ × ___ = ___ is _____ than ___			

8. Estimate by rounding both fractions to the nearest whole number (including 0).

a) $2\frac{1}{4} + 3\frac{5}{8} \approx 2 + 4 = 6$

b) $3\frac{1}{4} - 1\frac{4}{5} \approx$

c) $\frac{1}{5} + 4\frac{7}{9} \approx$

d) $2\frac{1}{3} \times 3\frac{5}{6} \approx$

e) $4\frac{4}{5} \times 3\frac{6}{7} \approx$

f) $6\frac{1}{7} \div 2\frac{2}{5} \approx$

g) $2\frac{3}{7} - \frac{1}{10} \approx$

h) $3\frac{1}{4} \div \frac{8}{9} \approx$

i) $5\frac{2}{3} + 1\frac{1}{4} \approx$

j) $4\frac{2}{5} - 1\frac{1}{3} \approx$

9. Kim wants to pour $\frac{3}{5}$ L from one container and $\frac{4}{9}$ L from another container into a third container that is 1 L size. Will the container overflow?

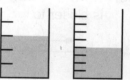

10. Scott's house is $\frac{3}{4}$ of a mile from school. After he rode $\frac{1}{10}$ of a mile on his bicycle, the tire went flat and he had to walk the rest of the way to school.

a) Estimate: Did he walk more or less than $\frac{1}{2}$ mile? Explain how you know.

b) Find the actual distance that he walked.

11. a) Is $\frac{13}{14} - \frac{5}{7}$ more or less than $\frac{1}{2}$? Explain.

b) Is $1\frac{3}{7} - \frac{1}{2}$ more or less than $\frac{1}{2}$? Explain.

c) Find the exact answers in parts a) and b).

12. A recipe calls for $2\frac{1}{4}$ cups of blueberries. Another recipe calls for $1\frac{5}{8}$ cups.

a) About how many cups of blueberries are needed for both recipes?

b) Find the exact answer in part a).

13. Tom has played soccer for $3\frac{5}{6}$ years. Mike has played soccer for $5\frac{1}{12}$ years. Is it reasonable to say that Mike has played soccer 2 years longer than Tom? Explain.

1. $\frac{2}{100}$ of Antarctica is *not* covered in ice.
What fraction of Antarctica is covered in ice?

2. Mia runs around a field 6 times in a $\frac{1}{2}$ hour.

The distance around the field is 0.5 miles.

How far can she run in an hour?

3. Seven classes at Washington Elementary School are going skiing.

There are 22 students in each class.

The teachers order 6 buses, which each hold 26 students. Will there be enough room? Explain.

4. It takes Kim 40 minutes to finish her homework: she spends $\frac{2}{5}$ of the time on math and $\frac{2}{5}$ of the time on science.

a) How many minutes did she spend on math and science?

b) How many minutes did she spend on other subjects?

c) What fraction of the time did she spend on other subjects?

5. Eric bikes $11\frac{7}{8}$ miles on Saturday.
He bikes $3\frac{1}{4}$ fewer miles on Sunday.

a) About how many miles did Eric bike on Sunday?

b) About how many miles did he bike in total?

6. Ron buys $\frac{8}{3}$ cups of sugar. He uses $\frac{1}{4}$ of it to bake muffins. He then eats $\frac{1}{6}$ of the muffins.

How much sugar did he eat?

7. A ball is dropped from a height of 24 m.
Each time it hits the ground, it bounces $\frac{3}{4}$ of the last height. How high did it bounce …

a) on the first bounce?

b) on the second bounce?

8. John earns $16.54 on Monday and adds it to his savings.

On Friday, he spends half of his money on a T-shirt. He now has $14.37.

How much money did he have before he started work Monday?

9. A pentagonal box has a perimeter of $3\frac{3}{4}$ m. How long is each side?

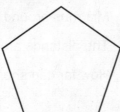

10. Roberto, Kendra, and Nathan paint a whole wall. Roberto paints $\frac{2}{5}$ of the wall and Kendra paints $\frac{1}{3}$.

Roberto	Kendra	Nathan
$\frac{2}{5}$	$\frac{1}{3}$	?

←————— 30 m —————→

a) What fraction of the wall did Nathan paint?

b) Each person paints a rectangular section. The wall is 30 m long. How long is each person's section?

11. Find the mystery numbers.

a) I am a number between 25 and 35. I am a multiple of 3 and 5.

b) I am a number between 10 and 20. My tens digit is 1 less than my ones digit.

c) Rounded to the nearest tens, I am 40. I am an even number. The difference in my digits is 3.

12. Tom gives away $\frac{3}{4}$ of his hockey cards.

a) What fraction of his cards did he keep?

b) Tom puts his remaining cards in a scrapbook. Each page holds 14 cards. He fills 23 pages. How many cards did he put in the book?

c) How many cards did he have before he gave away part of his collection?

13. Tony buys a binder for $17.25 and a pen for $2.75. He pays $\frac{7}{100}$ of the total price in taxes. How much did he pay in taxes?

14. Blanca spends $500.00 on furniture. She spends $\frac{3}{10}$ of the money on a chair, $50.00 on a table, and the rest on a sofa.

What fraction of the $500.00 did she spend on each item?

15. A leap year happens every 4 years. How many times will a leap year happen in 60 years?

16. The heart pumps about $\frac{1}{16}$ L of blood with each beat. About how many times would the heart need to beat to pump 4 L of blood?

17. The price of a movie ticket is $7.00 now. If the price rises by $0.15 each year, how much will the ticket cost in 10 years?

18. A cup of raspberries weighs about 7 ounces. A recipe calls for $2\frac{1}{4}$ cups of raspberries. About how many ounces of raspberries are needed?

OA5-6 Order of Operations and Brackets

Add and subtract in the order you read: from left to right.

1. Add or subtract from left to right.

a) $5 + 4 - 3$
$= 9 - 3$
$= 6$

b) $6 - 4 + 1$

c) $4 + 5 + 3$

d) $9 - 3 - 2$

Multiply and divide in the order you read: from left to right.

2. Multiply or divide from left to right.

a) $3 \times 4 \div 2$
$= 12 \div 2$
$= 6$

b) $6 \div 3 \times 2$

c) $7 \times 3 \times 2$

d) $12 \div 3 \div 2$

When doing operations…

Step 1: Do all multiplications and divisions from left to right.

Step 2: Do all additions and subtractions from left to right.

3. Circle the operation you would do first.

a) $5 + (2 \times 3)$

b) $9 - 2 + 5$

c) $10 + 5 \div 5$

d) $11 - 8 \div 2$

e) $12 \div 3 \times 2$

f) $10 - 3 \times 3$

g) $8 + 2 - 5$

h) $5 \times 5 - 6$

i) $18 \div 6 + 3$

j) $15 \div 5 - 2$

k) $2 \times 3 + 4$

l) $4 \times 6 \div 2$

4. Circle and do the first operation. Then rewrite the rest of the expression.

a) $(5 + 8) - 4$
$= \underline{\quad 13 - 4 \quad}$

b) $5 + (6 \div 3)$
$= \underline{\quad 5 + 2 \quad}$

c) $12 \div 4 + 2$
$= \underline{\qquad\qquad}$

d) $18 \div 6 \times 3$
$= \underline{\qquad\qquad}$

e) $10 - 5 - 3$
$= \underline{\qquad\qquad}$

f) $2 \times 6 \div 3$
$= \underline{\qquad\qquad}$

g) $16 \div 4 - 3$
$= \underline{\qquad\qquad}$

h) $11 - 5 + 5$
$= \underline{\qquad\qquad}$

i) $2 \times 30 \div 20$
$= \underline{\qquad\qquad}$

j) $7 \times 4 - 3$
$= \underline{\qquad\qquad}$

k) $36 \div 4 + 3$
$= \underline{\qquad\qquad}$

l) $20 - 5 \times 3$
$= \underline{\qquad\qquad}$

Brackets change the order of operations. Do the operations in brackets before all others.

Example: $7 - 3 + 2 = 4 + 2$ but $7 - (3 + 2) = 7 - 5$

$= 6$ $= 2$

5. Do the operation in brackets first. Then write the answer.

a) $(7 + 3) \times 2$

$= 10 \times 2$

$= 20$

b) $7 + (3 \times 2)$

c) $(7 + 3) \div 2$

d) $(7 - 3) \div 2$

e) $7 - (3 \times 2)$

f) $(7 - 3) \times 2$

g) $2 + (3 - 1)$

h) $8 - (6 \div 3)$

i) $4 \times (2 \times 3)$

j) $(4 \times 2) \times 3$

k) $(12 \div 6) \div 2$

l) $12 \div (6 \div 2)$

6. a) Add the same numbers in two ways. Do the addition in brackets first.

i) $(2 + 3) + 8$ $2 + (3 + 8)$

$= \underline{\quad} + 8$ $= 2 + \underline{\quad}$

$= \underline{\quad}$ $= \underline{\quad}$

ii) $(5 + 2) + 4$ $5 + (2 + 4)$

$= \underline{\quad} + \underline{\quad}$ $= \underline{\quad} + \underline{\quad}$

$= \underline{\quad}$ $= \underline{\quad}$

b) Does the answer change depending on which addition you do first? _____

7. a) Subtract the same numbers in two ways. Do the subtraction in brackets first.

i) $(9 - 5) - 2$ $9 - (5 - 2)$

$= \underline{\quad} - \underline{\quad}$ $= \underline{\quad} - \underline{\quad}$

$= \underline{\quad}$ $= \underline{\quad}$

ii) $11 - (6 - 5)$ $(11 - 6) - 5$

$= \underline{\quad} - \underline{\quad}$ $= \underline{\quad} - \underline{\quad}$

$= \underline{\quad}$ $= \underline{\quad}$

b) Does the answer change depending on which subtraction you do first? _____

OA5-7 Numerical Expressions

A **numerical expression** is a combination of numbers, operation signs, and sometimes brackets that represents a quantity.

Example: These numerical expressions all represent 10.

$5 + 2 + 3$	$14 - 4$	$70 \div 7$	$(3 + 2) \times 2$

1. Calculate the numerical expression by considering the order of operations.

a) $2 + 5 + 1$ _____　　　b) 2×5 _____　　　c) $3 \times 2 \times 1$ _____

d) $3 + 4 \times 1$ _____　　　e) $3 + 4 \div 2$ _____　　　f) $8 \times 3 \div 2$ _____

g) $(1 + 3) \times 4$ _____　　　h) $3 + (6 \div 2)$ _____　　　i) $(6 \times 3) \div 2$ _____

j) $(10 - 4) \div 2$ _____　　　k) $10 - (4 \div 2)$ _____　　　l) $5 \times (3 \times 2)$ _____

2. Write the number 3 in the box and then calculate the expression.

a) $\boxed{3} + 4 \longrightarrow \underline{7}$　　　b) $\boxed{3} + 2 \longrightarrow$ _____　　　c) $\boxed{} + 5 \longrightarrow$ _____

d) $9 - \boxed{} \longrightarrow$ _____　　　e) $17 - \boxed{} \longrightarrow$ _____　　　f) $\boxed{} - 2 \longrightarrow$ _____

g) $2 \times \boxed{} \longrightarrow$ _____　　　h) $\boxed{} \times 5 \longrightarrow$ _____　　　i) $3 \times \boxed{} \longrightarrow$ _____

j) $6 \div \boxed{} \longrightarrow$ _____　　　k) $15 \div \boxed{} \longrightarrow$ _____　　　l) $\boxed{} \div 3 \longrightarrow$ _____

Any number can be in an expression—not just whole numbers.

Examples:

$2.7 + 4.1$	$\dfrac{4}{5} - \dfrac{1}{5}$	$(2 + 3) \div \dfrac{1}{4}$

3. Calculate the numerical expression.

a) $2.3 + 1.6$ _____　　　b) 3×2.1 _____　　　c) 2×3.2 _____

d) $\dfrac{2}{5} + \dfrac{1}{5}$ _____　　　e) $2 \div \dfrac{1}{3}$ _____　　　f) $3 \times \dfrac{2}{7}$ _____

g) $\left(\dfrac{1}{7} + \dfrac{3}{7}\right) \times 2$ _____　　　h) $5 + (4 \times 1.2)$ _____　　　BONUS ▶ $\left(\dfrac{1}{5} \times \dfrac{3}{4}\right) \times 2$ _____

An **equation** is a statement that has two equal expressions separated by an equal sign.

Examples: $14 - 4 = 70 \div 7$ $\quad$ $12 = 3 \times 4$

4. a) Circle two expressions in Question 1 that represent the same number.

   b) Write an equation using the two expressions.

   _____ = _____

5. Verify that the equation is true.

   a) $(4 + 3) \times 2 \ = \ (5 \times 3) - 1$

   $(4 + 3) \times 2$ and $(5 \times 3) - 1$

   $= 7 \times 2$ $\qquad\qquad = 15 - 1$

   $= 14$ $\qquad\qquad\quad = 14$

   b) $2 \times 4 \times 5 \ = \ 4 \times 10$

   $2 \times 4 \times 5$ and 4×10

   c) $3 + 11 \ = \ (3 + 1) + (11 - 1)$

   $3 + 11$ and $(3 + 1) + (11 - 1)$

   d) $3 + 11 \ = \ (3 + 2) + (11 - 2)$

   $3 + 11$ and $(3 + 2) + (11 - 2)$

6. Peter calculated $12 - 4 \times 2$ and got 16. What mistake did he make? Explain.

   _____

   _____

BONUS ▶ Add brackets to show which operation was done first.

   a) $8 - 3 \times 2 = 10$

   b) $11 - 3 \times 3 = 2$

   c) $7 \times 3 - 2 = 7$

   d) $12 \div 3 \times 2 = 2$

   e) $6 - 2 \times 3 = 0$

   f) $3 \times 6 \div 2 = 9$

OA5-8 Unknown Quantities and Equations

1. Some apples are inside a bag and some are outside the bag. The total number of apples is shown. Draw the missing apples in the bag.

a)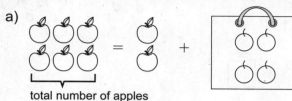

total number of apples

b)

c)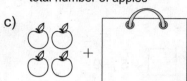

d)

2. Draw the missing apples in the bag. Then write an equation (with numbers) to represent the picture.

a)

    5 = 3 + ☐

b)

    ____ = ____ + ☐

c)

    ____ + ☐ = ____

d)

    ____ + ☐ = ____

3. Write an equation for each problem. Use a box for the unknown quantity.

a) There are 7 apples altogether. There are 4 outside a basket. How many are inside?

    7 = 4 + ☐

b) There are 9 apples altogether. There are 7 outside a basket. How many are inside?

    ____ = ____ + ☐

c) There are 11 plums altogether. There are 5 inside a bag. How many are outside?

d) 17 students are at the library. There are 9 in the computer room. How many are outside the computer room?

4. Jeff took some apples from a bag. Show how many apples were in the bag originally.

a) 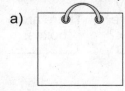 − 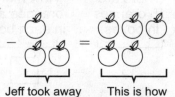 =

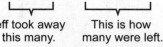

Jeff took away this many.   This is how many were left.

b)

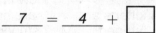

5. Show how many apples were in the bag originally. Then write an equation to represent the picture.

a)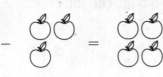

☐ − 3 = 4

b)

☐ − 2 = 6

6. Find the number that makes the equation true and write it in the box.

a) $\boxed{6} + 3 = 9$

b) $\boxed{} + 4 = 9$

c) $\boxed{} + 5 = 9$

d) $8 - \boxed{} = 5$

e) $13 - \boxed{} = 11$

f) $19 - \boxed{} = 8$

g) $3 \times \boxed{} = 6$

h) $4 \times \boxed{} = 20$

i) $2 \times \boxed{} = 2$

j) $\boxed{} \div 3 = 5$

k) $\boxed{} \div 5 = 3$

l) $\boxed{} \div 13 = 1$

m) $3 + 6 = 5 + \boxed{}$

n) $10 - 3 = \boxed{} + 4$

o) $1 + 5 = 7 - \boxed{}$

BONUS ▶ Put the same number in both boxes to make each equation true.

p) $\boxed{} + 3 = 11 - \boxed{}$

q) $\boxed{} \div 2 = \boxed{} - 2$

r) $\boxed{} \times \boxed{} = 16$

7. Find the missing number for the problem and write it in the box.

a) There are 10 marbles.   4 are outside the box.   How many are inside?

10 = 4 + ☐

b) There are 9 cards.   6 are outside the box.   How many are inside?

9 = 6 + ☐

c) There are 12 children in a class.   7 are girls.   How many are boys?

12 = 7 + ☐

d) A cat had 7 kittens.   4 kittens are striped.   How many are not striped?

7 = 4 + ☐

e) Paul had some stickers.   He gave away 3.   4 were left.

☐ − 3 = 4

f) There are 15 oranges in boxes.   How many oranges are in each box?   There are 3 boxes.

15 ÷ ☐ = 3

OA5-9 Translating Words into Expressions

1. Match the description with the correct numerical expression.

2 more than 6	4×6		2 divided into 11	3×11
6 divided by 3	$6 - 2$		11 reduced by 4	$11 \div 2$
2 less than 6	$6 + 2$		11 times 3	$11 + 3$
the product of 6 and 4	$6 - 3$		twice as many as 11	$11 - 4$
6 decreased by 3	$6 \div 3$		11 increased by 3	2×11

2. Write an expression for each description.

a) 4 more than 3 ___$3 + 4$___

b) 15 decreased by 8.5 _____

c) 24 divided by 8 _____

d) 2 less than 9 ___$9 - 2$___

e) 6.7 increased by 2.9 _____

f) 3.5 added to 4 _____

g) twice as many as 5 _____

h) 12 divided by 5 _____

i) the product of 7 and $\frac{2}{3}$ _____

j) 5 times $\frac{1}{3}$ _____

3. Turn the written instructions into mathematical expressions.

a) Add 8 and 3. ___$8 + 3$___

b) Divide 6 by 2. _____

c) Add 3.4 and 9. _____

d) Subtract 5 from 7. _____

e) Multiply 4.2 and 2. _____

f) Decrease 3 by $\frac{2}{5}$. _____

g) Add 8 and 4. Then divide by 3. _____

h) Divide 8 by 4. Then add 5. _____

i) Divide 4 by 2. Then add 10. Then subtract 4. _____

j) Multiply 6 and 5. Then subtract 20. Then divide by 2. _____

4. Write the mathematical expressions in words.

a) $(6 + 2) \times 3$ ___*Add 6 and 2. Then multiply by 3.*___

b) $(6 + 1) \times 2$ _____

c) $12 - 5 \times 2$ _____

d) $(3 - 2) \times 4$ _____

e) $4 \times (3 - 1 + 5)$ _____

5. How far will a motorcycle travel at the speed and in the time given? Write the numerical expression.

   a) Speed: 60 miles per hour
      Time: 2 hours

      Distance: ___60 × 2___ miles

   b) Speed: 80 km per hour
      Time: 4 hours

      Distance: _____ km

   c) Speed: 70 km per hour
      Time: 5 hours

      Distance: _____ km

6. a) How much will it cost to rent a bike for the time given? Write the numerical expression.

      i) 1 hour: ____5 × 1____   ii) 2 hours: _____   iii) 4 hours: _____

   b) Complete to explain the expression.

      i) 5 × 3 is the cost of renting a bike for __3__ hours.

      ii) 5 × 2 is the cost of renting a bike for ____ hours.

      iii) 5 × 5 is the cost of renting a bike for ____ hours.

   RENT A BIKE
   $5 an hour

7. a) A different rental company charges $6 per bike and then $3 for each hour.
      Write the numerical expression for the cost of renting a bike for…

      i) 1 hour: ___6 + 3 × 1___   ii) 2 hours: _____   iii) 4 hours: _____

   b) Complete the description of each expression.

      i) 6 + 3 × 3 is the cost of renting a bike for __3__ hours.

      ii) 6 + 3 × 2 is the cost of renting a bike for ____ hours.

      iii) 6 + 3 × 5 is the cost of renting a bike for ____ hours.

8. A field trip for a Grade 5 class costs $11 per student plus $2 for a snack.

   a) Write an expression to represent the cost for 1 student and 1 snack.   _____

   b) Write an expression to represent the cost for 3 students and 3 snacks.   _____

   BONUS ▶ Write a word problem that could be represented by 19 × (11 + 2).

9. A day pass can be used by 2 adults and 2 children for unlimited one-day bus travel on weekends. Write an expression to represent the number of day passes that are needed for 10 adults and 10 children.

10. 20 students from each class go to the museum. There are 5 classes, along with 13 teachers and 16 parents.

   a) Write an expression to represent the number of people that go to the museum.

   b) How many buses will be needed if 30 people ride in each bus?

Maria has 7 music albums. Clara has 3 times as many albums as Maria does. Maria uses a **tape diagram** to find out the number of albums Clara has.

Maria's albums: | 7 |

Clara's albums: | 7 | 7 | 7 | ⟵——— Clara has $3 \times 7 = 21$ albums.

A tape diagram has two strips on top of each other with the same unit, but different sizes.

1. Draw a tape diagram to model the story.

a) Clara has 3 stickers. Yu has 4 times as many stickers as Clara does.

Clara's stickers: | 3 |

Yu's stickers: | 3 | 3 | 3 | 3 |

b) There are 7 blue balloons. There are 3 times as many red balloons.

_____

_____

c) There are 8 red apples. There are 4 times as many green apples as red apples.

d) Blanca has 4 books. Mia has 5 times as many books.

You can use a tape diagram to find out the total number of something. Example:

Maria's albums: | 7 |

Clara's albums: | 7 | 7 | 7 | $7 + 21$ There are 28 albums in total.

2. Solve the problem by drawing a tape diagram.

a) Jin has 5 cards. Rob has 3 times as many cards as Jin. How many cards do they have together?

Jin's cards: 5 | 5 |

Rob's cards: 15 | 5 | 5 | 5 |

___5___ + ___15___ = ___20___ , so Jin and Rob have ___20___ cards together.

b) Luis studies toads and frogs. He has 6 toads and twice as many frogs. How many animals does he have altogether?

Toads: _____

Frogs: _____

_____ + _____ = _____ , so Luis has _____ animals altogether.

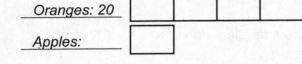

c) There are 24 round crackers in a box. There are 4 times as many square crackers in the box. How many crackers are there altogether?

d) There are 13 biographies in a school library. There are 3 times as many fiction books in the library. How many biographies and fiction books are in the library altogether?

3. Draw a tape diagram for the story. Then write the given number beside the correct bar.

a) There are 20 oranges. There are 4 times as many oranges as apples.

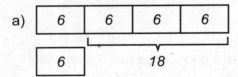

Oranges: 20

Apples:

b) There are 35 grandparents in the audience. There are 7 times as many grandparents as kids.

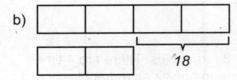

c) Layi spent $31.50 on her shoes and three times as much on her pants.

d) Grace studied math for 45 minutes and science for twice as much time.

4. All the blocks in each problem are the same size. What is the size of 1 block?

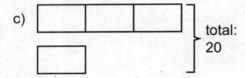

a)

| 6 | 6 | 6 | 6 |

| 6 | 18

b)

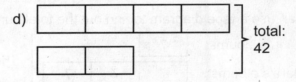

18

c)

total: 20

d)

total: 42

5. Show on the tape diagram what represents 12 beads. What is the size of 1 block?

a) There are 12 red beads.

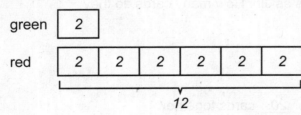

green | 2 |

red | 2 | 2 | 2 | 2 | 2 | 2 |

12

b) There are 12 beads in total.

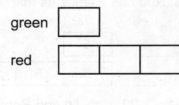

green

red

c) There are 12 more red beads than green.

green

red

d) There are 12 green beads.

green

red

OA5-11 Tape Diagrams II

1. The bars below represent the number of green (g) and red (r) beads in a box.
Fill in the blanks.

a) g: [|]

    r: [| |]

4 more red beads than green beads

1 block = _____ beads, so _____ beads in total

b) g: [| |]

    r: [ ]

12 beads altogether

1 block = _____ , so _____ green beads

c) g: [| |]

    r: [| | | |]

10 more red beads than green beads

1 block = _____ beads, so _____ beads in total

d) g: [| |]

    r: [| | |]

35 beads altogether

1 block = _____ , so _____ green beads

2. Use the tape diagram to find the number of red and green beads.

a) $\frac{2}{3}$ as many green beads as red beads

10 more red beads than green beads

g: [|]

r: [| |]

red: _____ green: _____

b) $\frac{3}{7}$ as many red beads as green beads.

30 beads altogether

r: [| |]

g: [| | | | | |]

red: _____ green: _____

c) $\frac{1}{4}$ as many green beads as red beads

12 more red beads than green beads

g: [ ]

r: [| | |]

red: _____ green: _____

d) $\frac{2}{5}$ as many red beads as green beads.

35 beads altogether

r: [|]

g: [| | | |]

red: _____ green: _____

3. Draw a tape diagram for the problem. Find the length of 1 block. Then solve the problem.

a) Jay has 3 times as many cards as Omar. Jay has 12 more cards than Omar. How many cards does each boy have?

Jay's cards	6	6	6

Omar's cards	6	12

Jay has __18__ cards, and Omar has __6__ cards.

b) Mike is 3 times as old as Kim. Mike is 14 years older than Kim. How old are Mike and Kim?

___Kim___

___Mike___

Mike is ___ years old, and Kim is ___ years old.

c) There are 6 times as many balloons as flags to decorate the house. There are 35 decorations altogether. How many balloons and how many flags are there?

d) To make 2 cups of boiled rice, you need 2 cups of water and 1 cup of dry rice. Anna wants to cook 4 cups of boiled rice. How much water and how much dry rice does she need?

There are _____ balloons and _____ flags.

Anna needs _____ cups of water and

_____ cups of dry rice.

4. Draw a model to answer the question.

Sara walks 4 times as far to school as Borana. Sara walks 6 more blocks than Borana. How many blocks do they each walk?

Sara:

Borana:

Sara: _____   Borana: _____

5. A pair of shoes costs twice as much as a T-shirt. George paid $34.50 for a pair of shoes and a T-shirt. How much does each item cost?

_____

_____

BONUS ▶ How much would George pay for three pairs of shoes and two T-shirts?

Operations and Algebraic Thinking 5-11

OA5-12 Variables

1. You can rent skates for $3 an hour. Write a numerical expression for the cost of renting skates for…

   a) 2 hours: __3 × 2__   b) 5 hours: _____   c) 6 hours: _____

> A **variable** is a letter or symbol (such as w, T, or h) that represents a number.
>
> To make an **algebraic expression**, replace some numbers in a numerical expression with variables.
>
> Examples of algebraic expressions: $w + 1$ $3 + 4 \times T$ $2 + t - 3 \times h$

2. Write an expression for the distance a car would travel at the speed and in the time given.

   a) Speed: 60 km per hour   b) Speed: 50 miles per hour   c) Speed: 70 km per hour

   Time: 3 hours   Time: 4 hours   Time: h hours

   Distance: _____ km   Distance: _____ miles   Distance: _____ km

> In the product of a number and a variable, the multiplication sign is usually dropped.
>
> $3 \times T$ can be written as $3T$ and $5 \times z$ can be written as $5z$.

3. Renting skis costs $5 an hour. Write a numerical expression for the cost of renting skis for…

   a) h hours: __5 × h__ or __5h__   b) t hours: _____ or _____

   c) x hours: _____ or _____   d) n hours: _____ or _____

RENT SKIS
$5 an hour

> When replacing a variable with a number, use brackets.
>
> Example: Replacing n with 7 in the expression $3n$ gives $3(7)$, which is another way to write 3×7.

4. Write the number 2 in the brackets and evaluate.

   a) $5(2) = $ __5 × 2__ = __10__   b) $3(\ \ ) = $ _____ = ____   c) $4(\ \ ) = $ _____ = ____

   d) $2(\ \ ) + 5$ e) $4(\ \ ) - 2$ f) $6(\ \ ) + 3$

5. Replace the variable with the given number and then evaluate.

   a) $2h + 5$, $h = 3$ b) $3n + 2$, $n = 5$ c) $4t - 1$, $t = 2$

   $2(3) + 5$

   $= 6 + 5$

   $= 11$

   d) $2m + 7$, $m = 6$ e) $8 - 2z$, $z = 3$ f) $7n - 6$, $n = 4$

> There are 4 times as many cats as dogs.
>
> $4 > 1$, so there are more cats than dogs.
>
> The number of cats is the *larger* quantity (L).
> The number of dogs is the *smaller* quantity (S).

> There are $\frac{1}{3}$ as many pears as bananas.
> $\frac{1}{3} < 1$, so there are more bananas than pears.
>
> The number of bananas is the *larger* quantity (L).
> The number of pears is the *smaller* quantity (S).

1. Fill in the table.

		Larger Quantity (L)	Smaller Quantity (S)
a)	There are 4 times as many plums as apples.	*plums*	*apples*
b)	There are 3 times as many dogs as cats.		
c)	There are $\frac{1}{4}$ as many eggs as nests.		
d)	There are half as many boys as girls.		
e)	Ron is twice as old as Maria.	*Ron's age*	
f)	A cat is $\frac{1}{5}$ as heavy as a dog.		

> The **scale factor** is a number you multiply a quantity or a length by to change size.
>
> If the scale factor is greater than 1,
> $L = $ scale factor $\times S$.
>
> Example: $L = 4 \times S$
>
> If the scale factor is smaller than 1,
> $S = $ scale factor $\times L$.
>
> Example: $S = \frac{1}{3} \times L$

2. L is the larger quantity and S is the smaller quantity. Circle the correct equations.

   $L = 4 \times S$ $S = 4 \times L$ $L = \frac{1}{5} \times S$ $S = \frac{1}{3} \times L$ $L = 2.5 \times S$

3. Write L above the larger quantity and S above the smaller quantity. Then write the equation.

   L S
   a) A book is 3 times as heavy as a notebook. $\underline{\quad L = 3 \times S \quad}$

   b) A shelf is 3 times as tall as a stool. $\underline{\qquad\qquad}$

   c) My cat is half as heavy as my dog. $\underline{\qquad\qquad}$

   d) Alicia is twice as old as Yu. $\underline{\qquad\qquad}$

   e) There are $\frac{1}{5}$ as many mice as hamsters in a pet store. $\underline{\qquad\qquad}$

4. Write *L* above the larger quantity and *S* above the smaller quantity everywhere. Write the equation. Then replace the correct letter with the given number.

a) $\overset{L}{\text{A book}}$ is 4 times as heavy as a $\overset{S}{\text{notebook}}$.

   The book weighs 400 g.

   $\underline{\quad L = 4 \times S \quad}$

   $\underline{\quad 400 = 4 \times S \quad}$

b) There are 3 times as many pears as apples.

   There are 12 apples.

   $\underline{\qquad\qquad\qquad}$

   $\underline{\qquad\qquad\qquad}$

c) Ava is half as old as Ken.

   Ken is 6 years old.

   $\underline{\qquad\qquad\qquad}$

   $\underline{\qquad\qquad\qquad}$

d) A tree is 6 times as tall as a bush.

   The tree is 18 m tall.

   $\underline{\qquad\qquad\qquad}$

   $\underline{\qquad\qquad\qquad}$

e) Mona has $\frac{1}{5}$ as much money as Leo.
   Leo has $10.

   $\underline{\qquad\qquad\qquad}$

   $\underline{\qquad\qquad\qquad}$

f) A break is $\frac{1}{6}$ as long as a lesson.
   The break lasts 15 minutes.

   $\underline{\qquad\qquad\qquad}$

   $\underline{\qquad\qquad\qquad}$

The equations $L = 4 \times S$ and $S = \frac{1}{4} \times L$ mean the same thing.

---

5. Write the equation that means the same thing.

a) $L = 2 \times S$

   $S = \frac{1}{2} \times L$

   $\underline{\qquad\qquad}$

b) $S = \frac{1}{5} \times L$

   $\underline{\qquad\qquad}$

c) $L = 6 \times S$

   $\underline{\qquad\qquad}$

d) $S = \frac{1}{3} \times L$

   $\underline{\qquad\qquad}$

6. Write the equation and replace the correct letter with the given number.
   If the unknown is not by itself, write the equation that means the same thing.
   Solve the equation to solve the problem.

a) There are 3 times as many cats as dogs in a shelter. There are 12 cats in the shelter. How many dogs are there?

b) Nina is half as old as Jose. Jose is 12. How old is Nina?

c) A snake is 5 times as long as a lizard. The snake is 125 cm long. How long is the lizard?

d) A scarf costs $\frac{1}{3}$ as much as a hat. The scarf costs $8. How much does the hat cost?

NBT5-54 Multiplying Decimals by Whole Numbers

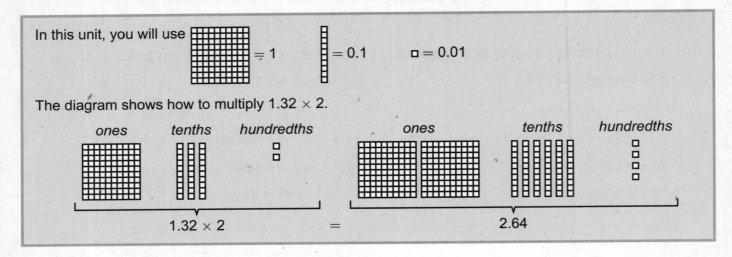

In this unit, you will use ▦ = 1   ▯ = 0.1   ▫ = 0.01

The diagram shows how to multiply 1.32 × 2.

| ones | tenths | hundredths | | ones | tenths | hundredths |

$$1.32 \times 2 \qquad = \qquad 2.64$$

1. Multiply by writing the decimals using tenths and hundredths.

a) 2.34 → __2__ ones + __3__ tenths + __4__ hundredths
    × 2 _____ × 2

    = __4__ ones + __6__ tenths + __8__ hundredths

    = _____4.68_____

b) 3.21 → _____ ones + _____ tenths + _____ hundredths
    × 3 _____ × 3

    = _____ ones + _____ tenths + _____ hundredths

    = _____

c) 1.21 → _____ ones + _____ tenths + _____ hundredths
    × 4 _____ × 4

    = _____ ones + _____ tenths + _____ hundredths

    = _____

2. Multiply each digit separately. Remember to include the decimal point.

a) 2.13 × 3 = _____   b) 3.41 × 2 = _____   c) 2.12 × 4 = _____   d) 4.23 × 2 = _____

e) 1.3 × 3 = _____   f) 2.01 × 4 = _____   g) 3.2 × 3 = _____   h) 3.01 × 3 = _____

3. A bowl of soup costs $2.12. Find the cost of 4 bowls.

4. Peanuts are on sale for $3.21 per pound. Find the cost of 3 pounds of peanuts.

BONUS ▶ Multiply mentally: 132,403.42 × 2 = _____

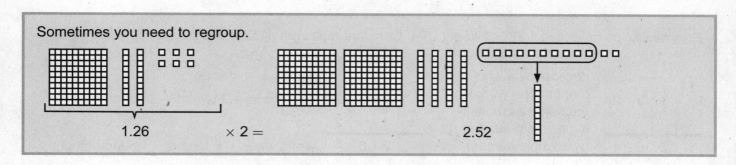

Sometimes you need to regroup.

1.26   × 2 =   2.52

5. Multiply by exchanging hundredths for tenths.

a) 2 × 3.17 = ___6___ ones + ___2___ tenths + ___14___ hundredths

    = ___6___ ones + ___3___ tenths + ___4___ hundredths = ___6.34___

b) 2 × 2.38 = _____ ones + _____ tenths + _____ hundredths

    = _____ ones + _____ tenths + _____ hundredths = _____

c) 3 × 2.15 = _____ ones + _____ tenths + _____ hundredths

    = _____ ones + _____ tenths + _____ hundredths = _____

6. Multiply by exchanging tenths for ones.

a) 2 × 4.63 = ___8___ ones + ___12___ tenths + ___6___ hundredths

    = ___9___ ones + ___2___ tenths + ___6___ hundredths = ___9.26___

b) 3 × 2.42 = _____ ones + _____ tenths + _____ hundredths

    = _____ ones + _____ tenths + _____ hundredths = _____

c) 4 × 2.31 = _____ ones + _____ tenths + _____ hundredths

    = _____ ones + _____ tenths + _____ hundredths = _____

7. Multiply by exchanging hundredths for tenths *or* tenths for ones.

a) 4 × 1.42 = _____ ones + _____ tenths + _____ hundredths

    = _____ ones + _____ tenths + _____ hundredths = _____

b) 3 × 2.17 = _____ ones + _____ tenths + _____ hundredths

    = _____ ones + _____ tenths + _____ hundredths = _____

c) 4 × 2.41 = _____ ones + _____ tenths + _____ hundredths

    = _____ ones + _____ tenths + _____ hundredths = _____

Previously, you multiplied 1.32 × 2 to get 2.64. Compare multiplying 132 × 2 and 1.32 × 2 using a grid.

When multiplying a decimal by a whole number, the work is the same as multiplying a whole number by a whole number.

1	.	3	2	← 1 one + 3 tenths + 2 hundredths
×			2	
2	.	6	4	← 2 ones + 6 tenths + 4 hundredths

8. Multiply using a grid. Remember to include the decimal point. You may have to regroup.

a)
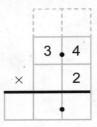

b)
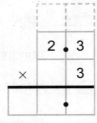

c)

d)
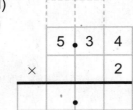

e)
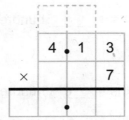

f)
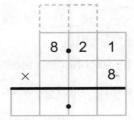

9. Multiply. Remember that 4 × 2.1 = 2.1 × 4.

a) 4 × 2.1   b) 3 × 4.2   c) 5 × 1.3   d) 7 × 2.3

e) 3 × 1.23   f) 4 × 2.03   g) 5 × 1.62   h) 9 × 2.36

i) 6 × 13.52   j) 7 × 23.15   k) 8 × 17.35   l) 9 × 41.23

REMINDER ▶ To multiply a decimal by 10, move the decimal point 1 place to the right.

Example: 3 . 2 5 × 10 = 3 2 . 5

You can rewrite 40 × 2.1 to make multiplying easier.

$$40 \times 2.1 = (4 \times 10) \times 2.1$$
$$= 4 \times (10 \times 2.1)$$
$$= 4 \times 21$$
$$= 84$$

10. Use the method above to multiply.

a) 30 × 4.1   b) 50 × 7.3   c) 60 × 3.58   d) 70 × 12.13

BONUS ▶ Find the product: 20 × 4,326.17

To multiply $\frac{3}{10} \times \frac{5}{10}$, find $\frac{3}{10}$ of $\frac{5}{10}$ Shade in 3 columns to show 3 tenths.

$$\frac{3}{10} \text{ of } \frac{5}{10} = \frac{15}{100}$$

Shade in 5 rows to show 5 tenths.

So $\frac{3}{10} \times \frac{5}{10} = \frac{15}{100}$

$\frac{15}{100}$

1. Write a multiplication equation for the diagram.

a)

b)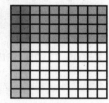

c)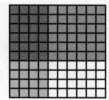

$$\frac{3}{10} \times \frac{4}{10} = \frac{12}{100}$$

_____   _____

2. Write the decimal fraction as a decimal.

a) $\frac{3}{10} =$ ___0.3___

b) $\frac{17}{100} =$ ___0.17___

c) $\frac{7}{10} =$ _____

d) $\frac{23}{10} =$ _____

e) $\frac{9}{10} =$ _____

f) $\frac{7}{100} =$ _____

g) $\frac{41}{10} =$ _____

h) $\frac{137}{100} =$ _____

3. Multiply the fractions.

a) $\frac{5}{10} \times \frac{3}{10} = \frac{15}{100}$

b) $\frac{7}{10} \times \frac{2}{100} = \frac{14}{1,000}$

c) $\frac{4}{10} \times \frac{9}{10} =$ _____

d) $\frac{8}{10} \times \frac{3}{100} =$ _____

e) $\frac{6}{10} \times \frac{7}{10} =$ _____

f) $\frac{5}{10} \times \frac{7}{100} =$ _____

BONUS ▶

g) $\frac{7}{10} \times \frac{3}{100} \times \frac{1}{10} =$ _____

h) $\frac{1}{100} \times \frac{2}{100} \times \frac{3}{100} =$ _____

4. Look at your answers to Question 3. How can you determine the number of zeros in the denominator of the product *before* multiplying?

_____

_____

5. Multiply the fractions. Then write the fraction equation as a decimal equation.

a) $\dfrac{6}{10} \times \dfrac{7}{10} = \dfrac{42}{100}$ b) $\dfrac{8}{10} \times \dfrac{3}{10} = $ _____   c) $\dfrac{7}{10} \times \dfrac{8}{100} = $ _____   d) $\dfrac{9}{10} \times \dfrac{6}{100} = $ _____

    $0.6 \times 0.7 = 0.42$ _____   _____   _____

6. Multiply the decimals as decimal fractions. Then write the answer as a decimal.

a) $0.3 \times 0.5 = \dfrac{3}{10} \times \dfrac{5}{10} = \dfrac{15}{100} = 0.15$ b) $0.7 \times 0.03 = \dfrac{7}{10} \times \dfrac{3}{100} = $

c) $0.7 \times 0.8 = $ d) $0.04 \times 0.5 = $

7. For each part in Question 6, find the number of zeros in the denominators and the number of times to shift the decimal digits.

a)

	$\dfrac{3}{10}$	$\times$	$\dfrac{5}{10}$	$=$	$\dfrac{15}{100}$
Number of zeros in denominator	*1*	$+$	*1*	$=$	*2*

	0.3	$\times$	0.5	$=$	0.15
Number of times to shift decimal point	*1*	$+$	*1*	$=$	*2*

b)

	$\dfrac{7}{10}$	$\times$	$\dfrac{3}{100}$	$=$	$\dfrac{21}{1,000}$
Number of zeros in denominator		$+$		$=$	

	0.7	$\times$	0.03	$=$	0.021
Number of times to shift decimal point		$+$		$=$	

c)

		$\times$		$=$	
Number of zeros in denominator		$+$		$=$	

		$\times$		$=$	
Number of times to shift decimal point		$+$		$=$	

d)

		$\times$		$=$	
Number of zeros in denominator		$+$		$=$	

		$\times$		$=$	
Number of times to shift decimal point		$+$		$=$	

8. What do you notice about the number of zeros in the denominator and the number of places the decimal point was shifted?

_____

NBT5-56 Multiplying Decimals by Decimals

Remember: When multiplying decimal fractions, you can add the number of zeros in each denominator to find the number of times to move the decimal point. In the same way, when multiplying decimals, you can add the number of decimal digits in each decimal.

Example:	Step 1	Step 2
31.2×0.04	Multiply the decimals as if they were whole numbers. $312 \times 4 = 1,248$	31.2 has 1 digit after the decimal point, and 0.04 has 2 digits after the decimal point. $1 + 2 = 3$ So shift the decimal point 3 places left. 1 2 4 8 So $31.2 \times 0.04 = 1.248$.

1. Without multiplying, find the number of digits after the decimal point in the product.

a)

	Number of decimal digits
2.13	2
× 0.4	1
Product	3

b)

	Number of decimal digits
7.3	1
× 0.9	1
Product	2

c)

	Number of decimal digits
13.2	1
× 2.7	1
Product	2

d)

	Number of decimal digits
4.75	2
× 9.8	1
Product	3

e)

	Number of decimal digits
0.43	2
× 0.7	1
Product	3

f)

	Number of decimal digits
23.1	1
× 1.25	2
Product	3

2. Multiply the decimals. Show the rough work in e) and f).

a) $0.4 \times 0.3 = \underline{0.12}$

| | | | 1 | 2 . |

b) $0.6 \times 0.08 = \underline{0.048}$

| | | | | 4 | 8 . |

c) $0.7 \times 0.5 = \underline{0.35}$

| | | | 3 | 5 . |

d) $0.09 \times 0.4 = \underline{.036}$

| | | 0 | 3 | 6 . |

e) $0.63 \times 0.4 = \underline{.252}$ Rough Work

| | | 2 | 5 2 . |

		6	3
×			4
	2	5	2

f) $3.2 \times 0.08 = \underline{.256}$ Rough Work

| | | 2 5 6 . |

		3	2
×			8
	2	5	6

3. Multiply as if the numbers were *whole* numbers. Add the number of decimal digits to find where the decimal point should be placed in the answer.

a)

```
      2 . 3
  ×   1 . 6
  ---------
      1 3 8
      2 3
  ---------
      3 . 6 8
```

b)
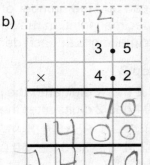

```
      3 . 5
  ×   4 . 2
  ---------
        7 0
    1 4 0 0
  ---------
    1 4 . 7 0
```

c)
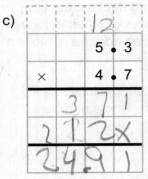

```
      5 . 3
  ×   4 . 7
  ---------
      3 7 1
    2 1 2 x
  ---------
    2 4 . 9 1
```

d)
```
      9 . 8
  ×   8 . 7
  ---------
    1 6 8 6
    7 8 4 x
  ---------
    8 5 . 2 6
```

4. Multiply.

a)

```
            1 2
            1 3
          6 . 3 7   ← 2 decimal digits
      ×     4 . 5   ← + 1 decimal digits
      -------------
637 × 5  →  3 1 8 5
637 × 40 → 2 5 4 8 0
      -------------
3,185 + 25,480 → 2 8 . 6 6 5   ← move decimal point
                                 3 places to the left
```

b)

```
      1 3 . 5
  ×     6 . 8
  -----------
```

c)

```
      1 . 2 9
  ×   0 . 4 8
  -----------
```

d)

```
      3 . 7 5
  ×   0 . 2 6
  -----------
```

e)

```
  6 2 0 1 2 . 9
  ×     0 . 3 5
  -------------
```

5. John earns $10.75 per hour at a fast-food restaurant. How much does he earn if he works 8.5 hours?

6. A pasta salad costs $4.95 per pound. Find the cost of 2.3 pounds of pasta salad.

1. A dime is 1.35 mm thick. How tall is a stack of 9 dimes?

2. There are 2.54 cm in 1 inch. A hockey puck is 3 inches wide. How wide is the puck in centimeters?

3. Tammy earns $11.25 per hour doing gardening. How much will she earn if she works 21 hours this week?

4. Fred's motorcycle gets 50.6 miles per gallon driving on the highway. His fuel tank holds 4.7 gallons. How far can he travel before having to fill up his tank of gas?

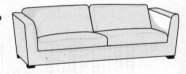

5. A wood screw costs 11.42¢. Dan needs a package of 400 screws.

  a) What is the cost of the screws in cents?

  b) What is the cost in dollars?

6. In Ventura County, CA, you multiply the price by 0.075 to find the sales tax.

  a) What is the sales tax charged when purchasing a couch priced at $530?

  b) What is the total cost including sales tax?

7. A quire of paper has 25 sheets and is 0.095 inches thick. A ream of paper has 20 quires.

  a) How many sheets of paper are in a ream?

  b) How thick is a ream of paper?

8. Steak costs $11.80 per pound at a supermarket. What is the cost of a steak weighing 0.75 pounds?

9. Light travels about 5.9 trillion miles in one year. This distance is called a *light year*.

  a) Find the distance in miles from the Sun to the following stars.

Star	Distance from the Sun (light years)	Distance from the Sun (trillions of miles)
Alpha Centauri	4.3	
Barnard's Star	6.0	
L726-8	8.7	
Sirius	9.5	
Ross 154	9.7	

  b) Which star is just over twice as far from the Sun as Alpha Centauri?

  c) How many light years further from the Sun is Sirius than L726-8?

REMINDER ▶ 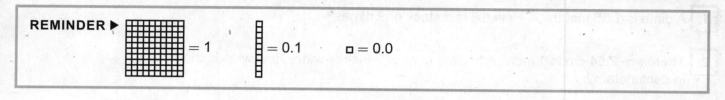 = 1 | = 0.1 □ = 0.0

1. Write the division equation for the base ten model.

a)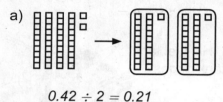

$0.42 \div 2 = 0.21$

b)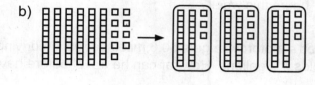

_____

c)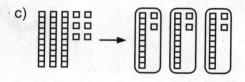

_____

d)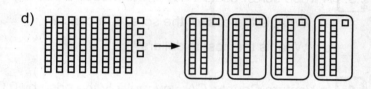

_____

e)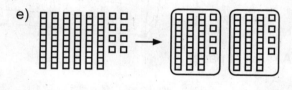

_____

f)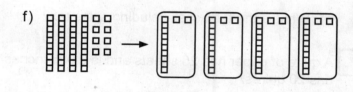

_____

2. Divide by writing the decimals using ones, tenths and hundredths.

a) $4.82 \div 2$

= (_4_ ones + _8_ tenths + _2_ hundredths) ÷ 2

= _2_ ones + _4_ tenths + _1_ hundredth

= _2.41_

b) $6.93 \div 3$

= (___ ones + ___ tenths + ___ hundredths) ÷ 3

= ___ ones + ___ tenths + ___ hundredth

= _____

c) $8.48 \div 4$

= (___ ones + ___ tenths + ___ hundredths) ÷ 4

= ___ ones + ___ tenth + ___ hundredths

= _____

d) $8.64 \div 2$

= (___ ones + ___ tenths + ___ hundredths) ÷ 2

= ___ ones + ___ tenths + ___ hundredths

= _____

BONUS ▶

e) $9.639 \div 3$

f) $8.048 \div 4$

3. Divide the decimal by a whole number by first dividing as if both numbers were whole numbers. Then count the number of decimal digits in the question to insert the decimal point in the answer.

a) $48 \div 2 = \underline{24}$

so $0.48 \div 2 = \underline{0.24}$

b) $63 \div 3 = \underline{}$

so $0.63 \div 3 = \underline{}$

c) $48 \div 4 = \underline{}$

so $0.48 \div 4 = \underline{}$

d) $246 \div 2 = \underline{123}$

so $2.46 \div 2 = \underline{1.23}$

e) $639 \div 3 = \underline{}$

so $6.39 \div 3 = \underline{}$

f) $488 \div 4 = \underline{}$

so $4.88 \div 4 = \underline{}$

Sometimes regrouping is required.

$1.26 \div 3 = (1 \text{ one} + 2 \text{ tenths} + 6 \text{ hundredths}) \div 3$

$ = (10 \text{ tenths} + 2 \text{ tenths} + 6 \text{ hundredths}) \div 3$

$ = (12 \text{ tenths} + 6 \text{ hundredths}) \div 3$

$ = 4 \text{ tenths} + 2 \text{ hundredths}$

$ = 0.42$

If we divide as if they were whole numbers, we get $126 \div 3 = 42$

```
        4   2
    ┌─────────
  3 │ 1   2   6
  − │ 1   2
    ├─────────
    │         6
  − │         6
    ├─────────
    │         0
```

---

4. The decimals have been divided as if they were whole numbers. Count the number of decimal digits to insert the decimal point.

a) $148 \div 2 = 74$

so $1.48 \div 2 = \underline{0.74}$

b) $216 \div 3 = 72$

so $2.16 \div 3 = \underline{}$

c) $364 \div 4 = 91$

so $3.64 \div 4 = \underline{}$

d) $156 \div 3 = 52$

so $15.6 \div 3 = \underline{}$

e) $328 \div 8 = 41$

so $32.8 \div 8 = \underline{}$

f) $459 \div 9 = 51$

so $45.9 \div 9 = \underline{}$

g) $105 \div 5 = 21$

so $1.05 \div 5 = \underline{}$

BONUS ▶ $24{,}608 \div 4 = 6{,}152$

so $24.608 \div 4 = \underline{}$

5. Tony walks 0.18 miles in 9 minutes. How far does he walk in 1 minute? $\underline{}$

6. Gilda earns \$36.93 in 3 hours. How much is she paid per hour? $\underline{}$

Dividing Decimals by Whole Numbers

REMINDER ▶ = 1 ‖ = 0.1 ▫ = 0.01

Example: This picture shows 5.34 with ones blocks, tenths blocks, and hundredths blocks.

1. Use the base ten models shown to find 5.34 ÷ 2. Complete your answer on the right.

Step 1: Divide the ones blocks into 2 equal groups.

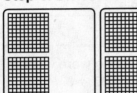 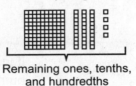

Remaining ones, tenths, and hundredths

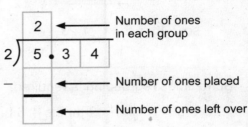

← Number of ones in each group

← Number of ones placed

← Number of ones left over

Step 2: Exchange the leftover ones block for 10 tenths blocks.

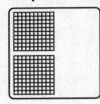  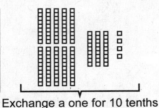

Exchange a one for 10 tenths

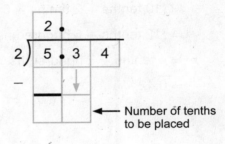

← Number of tenths to be placed

Step 3: Divide the tenths block into 2 equal groups.
Then exchange the leftover tenth block for 10 hundredths.

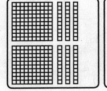 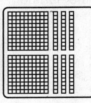 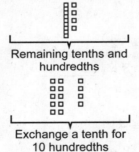

Remaining tenths and hundredths

Exchange a tenth for 10 hundredths

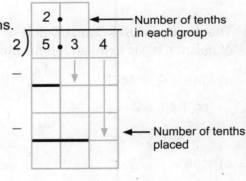

← Number of tenths in each group

← Number of tenths placed

Step 4: Divide the hundredths into the 2 equal groups.

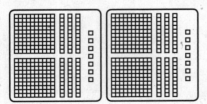

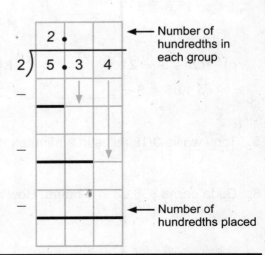

← Number of hundredths in each group

← Number of hundredths placed

So 5.34 ÷ 2 = _____

REMINDER ▶ 72 ÷ 12 = 6 ← quotient

dividend   divisor

To divide a decimal by a whole number:

Step 1: Divide as though they were whole numbers.

Step 2: Put the decimal point in the quotient directly above the decimal point in the dividend.

Example: 5.36 ÷ 4 = 1.34

	1	3	4
4)	5	3	6
−	4	↓	
	1	3	
−		1	2	↓
		1	6
−		1	6
			0

	1.	3	4
4)	5.	3	6
−	4	↓	
	1	3	
−		1	2	↓
		1	6
−		1	6
			0

2. Use 723 ÷ 3 = 241 to divide. Put the decimal point in the answer.

a)
	2	4	1
3)	7 •	2	3

b)
	2	4	1
3)	7	2 •	3

c)
	2	4	1
3)| 0 •| 7	2	3

d)
	2	4	1	
3)	7	2	3	0 •

3. Divide as though the decimal numbers are whole numbers. Then put the decimal point in the correct place.

a) 4) 7 • 3 2

b) 3) 5 1 • 6

c) 5) 0 • 6 8 5

d) 2) 1 3 7 • 0

e) 6)7.38

f) 7)94.5

g) 9)0.981

h) 8)196.0

4. A birthday gift costs $61.75. If the cost is split among 5 friends, how much does each friend pay?

5. A train traveled 291.5 miles in 5 hours. How many miles did it travel each hour?

6. A recipe for bread calls for 2 ounces of vegetable oil. How many loaves of bread can be made with a 33.8-ounce bottle of vegetable oil?

NBT5-60 Dividing Decimals by Decimals

> **REMINDER** ▶ A division statement can be written as a fraction. Example: $8 \div 4 = \dfrac{8}{4}$

1. Show how you can find an equivalent division statement by multiplying the numerator and denominator by 10.

   a) $8 \times 4 = \dfrac{8}{4} = \dfrac{8 \times 10}{4 \times 10} = (8 \times 10) \div (4 \times 10)$ b) $6 \div 2 =$

   c) $7 \div 8 =$ d) $4 \div 9 =$

> **REMINDER** ▶ If you multiply a decimal by 10, move the decimal point 1 place to the right.
>   If you multiply a decimal by 100, move the decimal point 2 places to the right.
>
> Example: $2.356 \times 10 = 23.56$ $2.356 \times 100 = 235.6$ $23 \times 10 = 23.0 \times 10 = 230$

2. Multiply mentally.

   a) $18.24 \times 10 =$ _____   b) $12.375 \times 100 =$ _____   c) $83 \times 100 =$ _____

   d) $2.193 \times 100 =$ _____   e) $0.34 \times 10 =$ _____   f) $0.013 \times 100 =$ _____

To divide decimals, you can multiply the dividend and divisor by the same number.
Multiply both numbers by the power of 10 that you need to make the divisor a whole number.

$0.27 \div 0.3 = (0.27 \times 10) \div (0.3 \times 10)$
$= 2.7 \div 3$
dividend   divisor $= 0.9 \longleftarrow$ quotient

---

3. Multiply the dividend and divisor by 10 or 100 to make the divisor a whole number. Then divide.

   a) $2.4 \div 0.4 = \underline{(2.4 \times 10) \div (0.4 \times 10)}$ b) $5.6 \div 0.07 =$ _____

   $= \underline{\quad 24 \div 4 \quad}$ $=$ _____

   $= \underline{\quad 6 \quad}$ $=$ _____

   c) $0.45 \div 0.05 =$ _____   d) $0.0056 \div 0.07 =$ _____

   $=$ _____ $=$ _____

   $=$ _____ $=$ _____

   e) $0.72 \div 0.8 =$ _____   f) $0.045 \div 0.09 =$ _____

   $=$ _____ $=$ _____

   $=$ _____ $=$ _____

To divide a decimal by a decimal, first multiply both the divisor and dividend by the power of 10 that you need to make the divisor a whole number.

Examples: 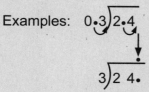   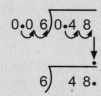

4. Multiply both the divisor and the dividend by the power of 10 that you need to make the divisor a whole number. Then put the decimal point for the quotient in the correct place. You may need to add zeros.

a) $0.4\overline{)\begin{array}{|c|c|c|} 1 & 3 & 2 \\ \hline 5 \cdot & 2 & 8 \end{array}}$

b) $0.4\overline{)\begin{array}{|c|c|c|} 1 & 3 & 2 \\ \hline 5 & 2 \cdot & 8 \end{array}}$

c) $0.04\overline{)\begin{array}{|c|c|c|} 1 & 3 & 2 \\ \hline 5 \cdot & 2 & 8 \end{array}}$

d) $0.04\overline{)\begin{array}{|c|c|c|c|} 1 & 3 & 2 & \\ \hline 5 & 2 \cdot & 8 & 0 \end{array}}$

e) $0.04\overline{)\begin{array}{|c|c|c|c|} 1 & 3 & 2 & \\ \hline 5 & 2 & 8 \cdot & 0 \end{array}}$

f) $0.004\overline{)\begin{array}{|c|c|c|} & 1 & 3 & 2 \\ \hline 0 \cdot & 5 & 2 & 8 \end{array}}$

5. Use long division to divide.

a) $5.28 \div 0.3$ $0.3\overline{)5.28}$

b) $9.48 \div 0.04$ $0.04\overline{)9.48}$

c) $49.6 \div 0.5$ $0.5\overline{)49.6}$

a)
$$3\overline{)\begin{array}{ccc} 5 & 2 \cdot & 8 \\ \hline \end{array}}$$
quotient: 1 . (with steps −3, 2)

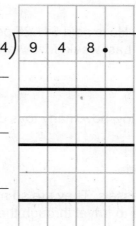

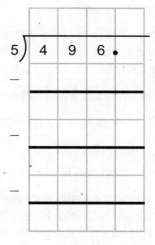

d) $1.35 \div 0.03$ e) $1.06 \div 0.4$ f) $3.46 \div 0.5$ g) $15.2 \div 0.02$

6. How many nickels are in $23.75?

7. A sunflower seed weighs approximately 0.4 g. How many seeds are in a 152 g bag?

1. A stationery store sells a box of paper clips for $1.92. If each paper clip costs $0.08, how many paper clips are in the box?

2. A race car track is in the shape of an oval. A car travels 1.6 km during one lap. How many laps of the track will a car have to make in order to travel 640 km?

3. A train travels 60.64 miles in 0.8 hours. How fast is it traveling each hour?

4. A bread recipe requires 0.3 ounces of baking powder. How many loaves of bread can be made with a 12-ounce can of baking powder?

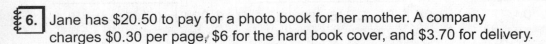

5. A patient needs 0.75 g of a medication per day. Each tablet has 0.5 g of medication. How many tablets should the patient take each day?

6. Jane has $20.50 to pay for a photo book for her mother. A company charges $0.30 per page, $6 for the hard book cover, and $3.70 for delivery.

   a) After paying for delivery and the cover, how much money is left to pay for the pages inside the book?

   b) How many pages can Jane include in her book?

7. Max works in a factory that makes toys. He is paid $11.25 per hour and is paid a bonus for each toy he produces. Last week, Max worked 35 hours and his total earnings were $450.

   a) How much was his bonus last week?

   b) If Max gets a $0.05 bonus for each toy, how many toys did he produce?

8. A red blood cell is about 0.000008 m wide.

   a) Capillaries are very narrow blood vessels. They are so narrow that red blood cells must line up in single file to travel through them. How many red blood cells will fit in a capillary 1 m long?

   b) Blood donors each give 1 pint of blood during a donation. A pint of blood contains about 2,400,000,000,000 red blood cells. If the cells were lined up end to end, how long would the line be?

capillary

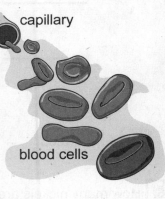

blood cells

9. If a patient in a hospital is dehydrated, he is sometimes given a saline solution intravenously. A saline solution is a mixture of salt and water, usually 0.009 g of salt for every 1 L of water. How many liters of solution were given to a patient who received 0.0135 g of salt?

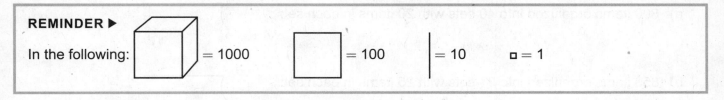

REMINDER ▶

In the following: = 1000   = 100   = 10   □ = 1

1. Write a division question for the diagram.

a)

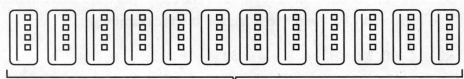

$$12\overline{)168}$$ with quotient 1 4

12 sets; 14 items in each set

b)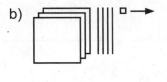

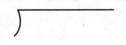

_____

c)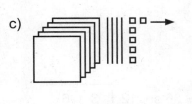

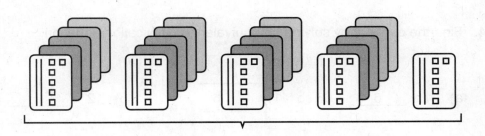

_____

d)

_____

2. Write a division equation for . . .

   a) 800 items organized into 40 sets with 20 items in each set:

   b) 850 items organized into 34 sets with 25 items in each set: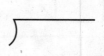

The division $4\overline{)32}$ leads to 4 facts in the **fact family:**

$32 \div 4 = 8,\ 32 \div 8 = 4,\ 8 \times 4 = 32,\ 4 \times 8 = 32$

---

3. Write the facts in the fact family.

   a) $15\overline{)195}$ (quotient 13) $195 \div 15 = 13$ $195 \div 13 = 15$ $13 \times 15 = 195$ $15 \times 13 = 195$

   b) $23\overline{)782}$ (quotient 34)   _____   _____   _____   _____

   c) $38\overline{)646}$ (quotient 17)   _____   _____   _____   _____

   d) $10\overline{)900}$ (quotient 90)   _____   _____   _____   _____

4. Find the quotient by solving an equivalent multiplication equation.

   a) $11 \times \underline{\ 6\ } = 66$ so $11\overline{)66}$

   b) $12 \times \underline{\hspace{1cm}} = 36$ so $12\overline{)36}$

   c) $22 \times \underline{\hspace{1cm}} = 88$ so $22\overline{)88}$

   d) $32 \times \underline{\hspace{1cm}} = 96$ so $32\overline{)96}$

   e) $42 \times \underline{\hspace{1cm}} = 84$ so $42\overline{)84}$

   f) $21 \times \underline{\hspace{1cm}} = 84$ so $21\overline{)84}$

   g) $25 \times \underline{\hspace{1cm}} = 100$ so $25\overline{)100}$

   h) $33 \times \underline{\hspace{1cm}} = 99$ so $33\overline{)99}$

   BONUS ▶

   $20 \times \underline{\hspace{1cm}} = 1,000$ so $20\overline{)1000}$

NBT5-63 2-Digit Division (Introduction)

> To find how many tens there are in a number, cover up the ones digit.
>
> Examples: 7 4 ⟶ 7X̶   There are 7 tens in 74.
>
>   3 5 8 ⟶ ⟨3 5⟩X̶   There are 35 tens in 358.
>
>   1 , 2 7 4 ⟶ ⟨1 , 2 7⟩X̶   There are 127 tens in 1,274.

1. Circle the number of tens.

    a) 3 2 5   b) 1 8 7   c) 9 4

    d) 7 6   e) 1, 2 8 3   f) 9, 5 6 4

2. Circle the number of tens in the dividend. Then say how many tens are being placed into how many groups.

    a) 12) 3 8 4   b) 52) 1 8 2 0   c) 71) 1 4 9 1

    __38__ tens in __12__ groups   _____ tens in _____ groups   _____ tens in _____ groups

    d) 39) 7 4 1   e) 36) 8 2 8   f) 24) 6 7 2

    _____ tens in _____ groups   _____ tens in _____ groups   _____ tens in _____ groups

> When rounding to the nearest ten, if the ones digit is 5 or higher, round *up* to the next multiple of 10.
>
> Examples: 5$\underline{7}$ ≈ 60   7 > 5 so round up   7$\underline{2}$ ≈ 70   2 < 5 so don't round up

3. Round each number to the nearest ten.

    a) 12 ≈ __10__   b) 36 ≈ _____   c) 24 ≈ _____   d) 39 ≈ _____

    e) 52 ≈ _____   f) 71 ≈ _____   g) 48 ≈ _____   h) 81 ≈ _____

4. Round the divisor to the nearest ten.

    a) ⟨12⟩) 3 8 4   b) 36) 8 2 8   c) 24) 7 6 8

        __10__   _____   _____

    d) 19) 4 7 6   e) 41) 9 4 3   f) 89) 1 6 9 1

    _____   _____   _____

Estimate the quotient using these steps.

Step 1: Round the divisor to the nearest ten. $48 \div 13$ *10, 20, 30, 40, 5̶0̶*

Step 2: Skip count by the rounded divisor. Stop just before you reach the dividend. $\approx 48 \div 10$

Step 3: The number of times you counted is an estimate for the quotient. ≈ 4

5. Estimate the quotient.

a) $68 \div 19$ _____

$\approx$

$\approx$

b) $185 \div 31$ _____

$\approx$

$\approx$

c) $123 \div 37$ _____

$\approx$

$\approx$

d) $208 \div 43$ _____

$\approx$

$\approx$

6. Round the divisor down to the nearest ten and circle the number of tens in the dividend. Skip count by the rounded divisor and stop before you pass the number of tens. Count the multiples.

a)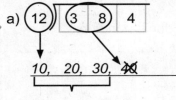

10, 20, 30, 4̶0̶

___*3*___ multiples

$\approx$ ___*3*___ tens in each group

b) 36) 8 2 8

_____

_____ multiples

$\approx$ ____ tens in each group

c) 24) 7 6 8

_____

_____ multiples

$\approx$ ____ tens in each group

d) 39) 7 4 1

_____

_____ multiples

$\approx$ ____ tens in each group

e) 52) 1 8 2 0

_____

_____ multiples

$\approx$ ____ tens in each group

f) 71) 1 4 9 1

_____

_____ multiples

$\approx$ ____ tens in each group

7. 18 people want to share $810. Approximately how many $10 bills can each person have?

Step 1: Estimate the first digit of the quotient by rounding the divisor to the nearest ten and skip counting until you pass the number of tens in the dividend.

Step 2: Use the number of multiples as the number of tens in the quotient. Multiply the first digit of the quotient by the divisor and subtract.

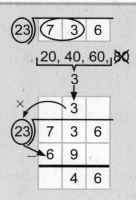

1. Perform the first two steps of long division with a two-digit divisor.

a) 31) 8 0 6

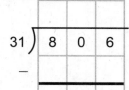

b) 38) 9 8 8

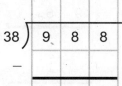

c) 17) 6 1 2

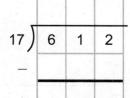

d) 41) 9 4 3

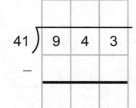

Step 3: Estimate the next digit of the quotient by rounding the divisor to the nearest ten and skip counting until you pass the number of **ones**.

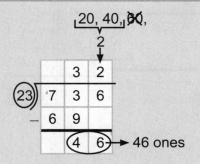

2. Skip count by the rounded divisor until you pass the number of ones.

a)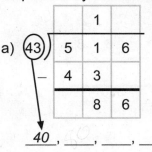

____, ____, ____, ____

____ multiple

b) 32) 7 3 6
− 6 4
9 6

____, ____, ____, ____

____ multiples

c) 12) 7 5 6
− 7 2
3 6

____, ____, ____, ____

____ multiples

Step 4: Multiply the ones digit of the quotient by the divisor and subtract.

```
×          3 2
23 ) 7  3  9
   − 6  9
     ────
        4  9
      − 4  6
        ────
           3
```

3. Complete the last step of the division.

a)
```
          3  1
32 ) 9  9  4
   − 9  6
     ──────
        3  4
   − 
     ──────
```

b)
```
          2  1
41 ) 8  6  5
   − 8  2
     ──────
        4  5
   − 
     ──────
```

c)
```
          4  2
18 ) 7  5  8
   − 7  2
     ──────
        3  8
   − 
     ──────
```

4. Divide. Use the grids to multiply for Steps 2 and 4.

a)
```
          2  6
11 ) 2  8  7
   − 2  2
     ──────
        6  7
      − 6  6
        ──────
           1
```
```
        1  1
×          2
   ──────
        2  2

        1  1
×          6
   ──────
     6  6
```

b)
```
          
56 ) 6  9  4
   − 
     ──────
   − 
     ──────
```
```
           5  6
×          
   ──────

           5  6
×          
   ──────
```

c)
```
          
43 ) 5  1  8
   − 
     ──────
   − 
     ──────
```
```
        4  3
×          
   ──────

        4  3
×          
   ──────
```

d)
```
          
68 ) 8  3  3
   − 
     ──────
   − 
     ──────
```
```
           6  8
×          
   ──────

           6  8
×          
   ──────
```

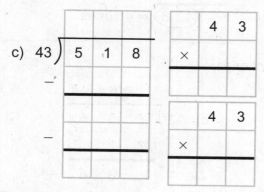
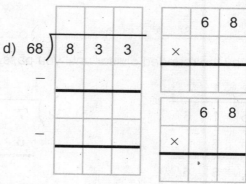

5. Jim wants to divide 851 ÷ 37. Explain how he should estimate the first digit of the quotient.

When estimating the quotient in a division problem, your estimate might turn out to be too high or too low.

Examples:

20 > 16 so
the estimate
is too low!

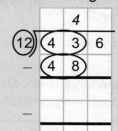

48 > 43 so
the estimate
is too high!

1. Is the estimate too high, too low, or just right?

a)
$$18 \overline{)572}$$
   $$-\ 36$$
   $$\ \ \ 21$$

_____ 21 > 18

_____ too low!

b)
$$23 \overline{)814}$$
   $$-\ 92$$

_____

_____

c)
$$28 \overline{)172}$$
   $$-\ 140$$
   $$\ \ \ \ 32$$

_____

_____

d)
$$72 \overline{)234}$$
   $$-\ 216$$
   $$\ \ \ \ 18$$

_____

_____

2. Use a first estimate to make a better estimate. Then perform the first two steps of division.

a)
$$37 \overline{)765}$$
   $$\ \ 37$$
   $$\ \ 39$$

too low!

b)
$$27 \overline{)531}$$
   $$\ \ 54$$

_____

c)
$$43 \overline{)854}$$
   $$\ \ 86$$

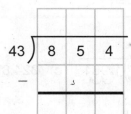

_____

d)
$$16 \overline{)973}$$
   $$\ \ 80$$

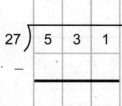

_____

BONUS ▶
$$32 \overline{)94,372}$$
   $$\ \ \ 96$$

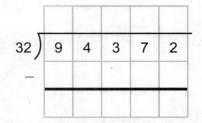

_____

3. Circle the first part of the dividend that is equal to or greater than the divisor.

divisor   dividend

a) 23)⟨5 8⟩6   b) 27)⟨1, 5 8⟩6   c) 72)9, 4 1 2   d) 52)3, 5 7 1   e) 68)6 5 7

  58 > 23   158 > 27

4. Circle the first part of the dividend that is equal to or greater than the divisor.
Then perform the first two steps of long division.

a) 32)⟨1, 9 5⟩2   b) 43)8 7 2   c) 46)1 3 8

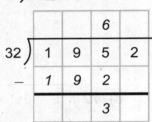

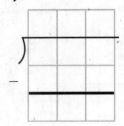

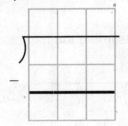

5. Divide using long division. Hint: Circle the first part of the dividend that is as least
as big as the divisor.

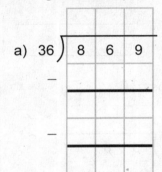

a) 36)8 6 9   b) 24)1 9 3   c) 18)1 1 3 4   d) 62)9 3 7

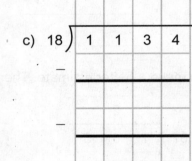

REMINDER ▶ When dividing by decimals, multiply the divisor and dividend by the same
power of 10 to make the divisor a whole number.

---

6. Divide using long division.

a) $9.43 \div 4.1$ 4.1)9.43   b) $1.961 \div 0.37$ 0.37)1.961   c) $324.8 \div 1.4$ 1.4)324.8

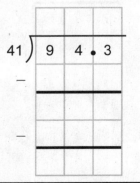

41)9 4 . 3

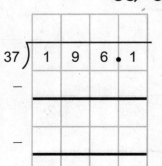

37)1 9 6 . 1

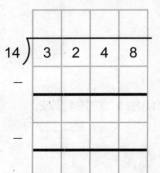

14)3 2 4 8

Number and Operations in Base Ten 5-65

1. Juice boxes are packaged 12 to a case. How many cases are needed for 432 juice bottles?

2. A parking garage collected $1,092 in parking fees for 78 cars. How much was the charge for each car?

3. A school takes 1,392 students to a theme park for a school trip. Each bus holds 24 students. How many buses were used?

4. A photo lab charges $0.12 per print. Ben has $7.80 to buy prints of his favorite photo. How many prints can he buy?

5. A cow produces about 30.3 L of milk a day. Farmer Jones has 50 cows. A carton of milk holds about 1.5 L of milk.

  a) How many liters of milk are produced by all of the cows?

  b) How many cartons are needed to hold all the milk?

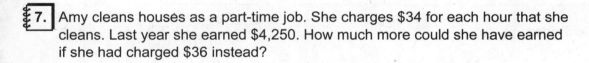

6. The speed of sound at sea level is 1,225,044.0 m per hour.

  a) How far does sound travel in 1 minute?

  b) How far does sound travel in 1 second?

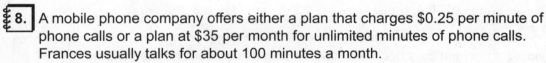

7. Amy cleans houses as a part-time job. She charges $34 for each hour that she cleans. Last year she earned $4,250. How much more could she have earned if she had charged $36 instead?

8. A mobile phone company offers either a plan that charges $0.25 per minute of phone calls or a plan at $35 per month for unlimited minutes of phone calls. Frances usually talks for about 100 minutes a month.

  a) What would she be charged if she chose the first plan?

  b) How many minutes of phone calls would Frances have to make in order to make the unlimited plan the better choice?

9. A banquet hall charges $160 in rent for an event, plus $25.60 for food per person attending.

  a) Find the total cost for 40 people.

  b) What should each person pay to cover the total costs?

> **REMINDER ▶** When multiplying decimals, multiply as if they were whole numbers. To find where to put the decimal point, find the number of decimal digits in each decimal.
>
> Example: 0.04×0.9 ⟶ multiply $4 \times 9 = 36$ so $0.04 \times 0.9 = 0.036$
>
> 2 decimal + 1 decimal   decimal point is moved
>   digits   digit   left 3 places

1. Use mental math to multiply.

a) $0.7 \times 0.5 = \underline{\ 0.35\ }$ b) $0.6 \times 0.4 = \underline{\hspace{2cm}}$ c) $0.2 \times 0.7 = \underline{\hspace{2cm}}$

d) $0.03 \times 0.9 = \underline{\hspace{2cm}}$ e) $0.5 \times 0.08 = \underline{\hspace{2cm}}$ f) $0.03 \times 0.04 = \underline{\hspace{2cm}}$

g) $0.6 \times 0.005 = \underline{\hspace{2cm}}$ h) $0.7 \times 0.3 = \underline{\hspace{2cm}}$ i) $0.9 \times 0.7 = \underline{\hspace{2cm}}$

2. Use mental math to multiply.

a) $2.3 \times 0.2 = \underline{\ 0.46\ }$ b) $3.2 \times 0.3 = \underline{\hspace{2cm}}$ c) $0.21 \times 4 = \underline{\hspace{2cm}}$

d) $4.2 \times 0.2 = \underline{\hspace{2cm}}$ e) $1.1 \times 0.09 = \underline{\hspace{2cm}}$ f) $0.31 \times 3 = \underline{\hspace{2cm}}$

> **REMINDER ▶** When dividing a decimal by a whole number, divide as if both numbers were whole numbers. Place the decimal point in the quotient directly above the decimal point in the dividend.

3. Use mental math to divide.

a) $7.2 \div 8 = \underline{\ 0.9\ }$ b) $4.2 \div 7 = \underline{\hspace{2cm}}$ c) $3.6 \div 9 = \underline{\hspace{2cm}}$

d) $3.5 \div 7 = \underline{\hspace{2cm}}$ e) $6.4 \div 8 = \underline{\hspace{2cm}}$ f) $6.3 \div 7 = \underline{\hspace{2cm}}$

> **REMINDER ▶** When dividing a decimal by a decimal, multiply both the divisor and the dividend by the same power of 10. This will make the divisor a whole number.
> Example: $0.32 \div 0.4 = (0.32 \times 10) \div (0.4 \times 10) = 3.2 \div 4 = 0.8$

4. Use mental math to divide.

a) $0.48 \div 0.6 = \underline{\ 0.8\ }$ b) $5.6 \div 0.7 = \underline{\hspace{2cm}}$ c) $0.49 \div 0.7 = \underline{\hspace{2cm}}$

d) $0.35 \div 0.07 = \underline{\hspace{2cm}}$ e) $0.025 \div 0.5 = \underline{\hspace{2cm}}$ f) $2.8 \div 0.007 = \underline{\hspace{2cm}}$

BONUS ▶ $0.000063 \div 0.0007 = \underline{\hspace{3cm}}$

Remember:

If you multiply by a fraction *larger* than 1, the answer will be *larger* than the number.

If you multiply by a fraction *smaller* than 1, the answer will be *smaller* than the number.

A decimal is just a fraction, so the rules apply to decimals as well.

Example: $6.3 \times 1.4 = 8.82$ and $8.82 > 6.3$ Example: $6.3 \times 0.82 = 5.166$ and $5.166 < 6.3$

5. Circle **T** for True or **F** for False.

a) $23.5 \times 0.3 > 23.5$ **T** (**F**)

b) $18.2 \times 1.7 > 18.2$ **T** **F**

c) $12.1 \times 0.08 < 12.1$ **T** **F**

d) $13.8 \times 1.25 < 13.8$ **T** **F**

e) $0.05 \times 1.3 > 0.05$ **T** **F**

f) $0.9 \times 0.2 < 0.9$ **T** **F**

g) $27.2 \times 1.25 < 27.2$ **T** **F**

h) $0.05 \times 0.99 < 0.05$ **T** **F**

i) $0.8 \times 1.01 < 0.8$ **T** **F**

j) $123 \times 0.85 > 123$ **T** **F**

If you divide by a decimal *larger* than 1.0, the answer will be *smaller* than the original number.

Example: $7.7 \div 1.1 = 7$ and $7 < 7.7$

If you divide by a decimal *smaller* than 1.0, the answer will be *larger* than the original number.

Example: $8.0 \div 0.8 = 10$ and $10 > 8.0$

6. Circle **T** for True or **F** for False.

a) $1.23 \div 0.6 > 1.23$ (**T**) **F**

b) $8.05 \div 2.3 > 8.05$ **T** **F**

c) $1.28 \div 0.4 < 1.28$ **T** **F**

d) $13.23 \div 2.1 < 13.23$ **T** **F**

e) $0.06 \div 0.2 < 0.06$ **T** **F**

f) $0.1 \div 2.0 > 0.1$ **T** **F**

g) $42.6 \div 1.1 < 42.6$ **T** **F**

h) $0.95 \div 0.8 < 0.95$ **T** **F**

i) $19.3 \div 1.95 > 19.3$ **T** **F**

j) $125 \div 0.9 > 125$ **T** **F**

REMINDER ▶ *Multiplying* a decimal by a power of 10 moves the decimal point to the *right*. *Dividing* a decimal by a power of 10 moves the decimal point to the *left*. The decimal point moves the same number of places as the number of zeros in the power of 10.

7. Multiply or divide using mental math.

a) $0.234 \times 100 =$ _____

b) $35.67 \div 10 =$ _____

c) $2.1876 \times 1,000 =$ _____

d) $12.57 \div 100 =$ _____

e) $0.76 \times 1,000 =$ _____

f) $7.875 \div 100 =$ _____

g) $9.234 \times 10 =$ _____

h) $7.68 \div 1,000 =$ _____

i) $0.0042 \times 100 =$ _____

When multiplying decimals, you can estimate the answer by rounding one of the numbers to 1, 10, or 100.

Examples: 0.95×43.2 $0.95 \approx 1$ 8.9×132.47 $8.9 \approx 10$

$\approx 1 \times 43.2$ $\approx 10 \times 132.47$

≈ 43.2 $\approx 1,324.7$ Move the decimal point right 1 place.

8. Estimate by rounding one of the numbers to the nearest power of 10. Then circle the correct answer from the list.

   a) 89.2×0.075 (0.669, (6.69), 66.9)

   $\approx$ _100 $\times$ 0.075_

   $\approx$ _7.5_

   b) 45.2×0.85 (3.842, 38.42, 384.2)

   $\approx$ _____

   $\approx$ _____

   c) 950×0.25 (23.75, 237.5, 2375)

   $\approx$ _____

   $\approx$ _____

   d) 1.35×25.6 (3.456, 34.56, 345.6)

   $\approx$ _____

   $\approx$ _____

When dividing decimals, it helps to use mental math to make sure you put the decimal point in the correct place. If one of the factors can be rounded to 1, 10, 100, or 1000, mental math can help you estimate.

Examples: $243.6 \div 0.87$ $0.87 \approx 1$ $21.85 \div 9.5$ $9.5 \approx 10$

$\approx 243.6 \div 1$ $\approx 21.85 \div 10$

≈ 243.6 ≈ 2.185 Move the decimal point left 1 place.

9. Use mental math to circle the correct answer from the list.

   a) $370.5 \div 11.4$ (0.325, (32.5), 325)

   $\approx$ _370.5 $\div$ 10_

   $\approx$ _37.05_

   b) $226.5 \div 98.5$ (0.023, 0.23, 2.3)

   $\approx$ _____

   $\approx$ _____

   c) $4.26 \div 0.75$ (5.68, 56.8, 568)

   $\approx$ _____

   $\approx$ _____

   d) $253.75 \div 0.875$ (2.9, 29, 290)

   $\approx$ _____

   $\approx$ _____

MD5-10 Inches

In the United States, we often measure length, height, and thickness in **inches.** We write **1 in** for 1 inch.

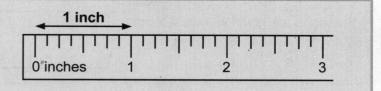

1. Measure the length of the line segment or object.

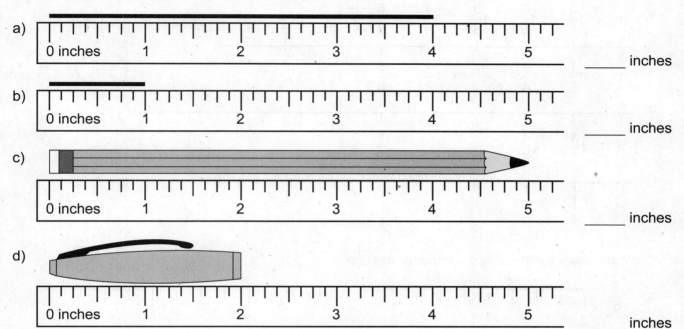

a)

_____ inches

b)

_____ inches

c)

_____ inches

d)

_____ inches

2. a) Measure all the sides of each shape.

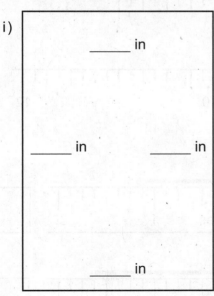

i)

_____ in

_____ in   _____ in

_____ in

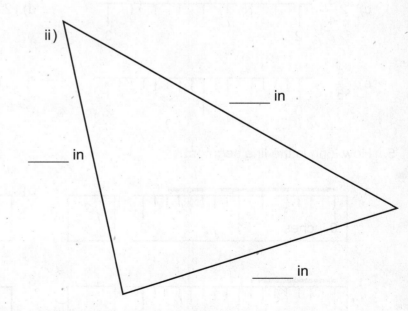

ii)

_____ in

_____ in

_____ in

_____ in

b) **Perimeter** is the distance around the shape. Find the perimeter of the shapes in part a).

i) Perimeter = _____ in   ii) Perimeter = _____ in

On some rulers, an inch is divided into 8 equal parts. This ruler is enlarged to show the fractions.

Fractions are given in lowest terms:
$\frac{2}{8} = \frac{1}{4}$, $\frac{4}{8} = \frac{1}{2}$, $\frac{6}{8} = \frac{3}{4}$

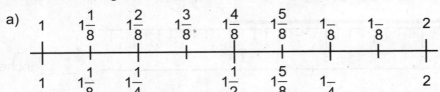

3. Write the missing mixed numbers on the number line.

a)

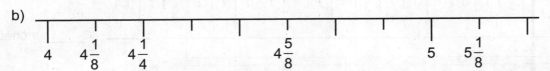

b)

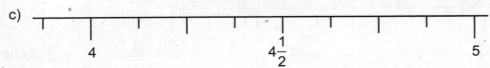

c)

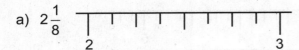

4. Circle the mark for the mixed number or fraction.

a) $2\frac{1}{8}$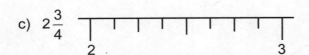

b) $7\frac{5}{8}$

c) $2\frac{3}{4}$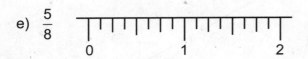

d) $2\frac{3}{8}$

e) $\frac{5}{8}$

f) $11\frac{1}{2}$

5. How long is the line segment?

a) 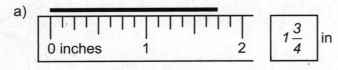 $1\frac{3}{4}$ in

b) 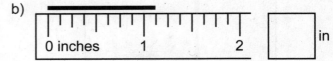 in

c) 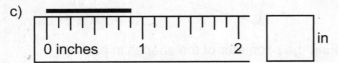 in

d) 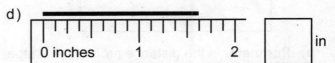 in

6. Measure the object. Use your ruler for parts d) to g).

a) 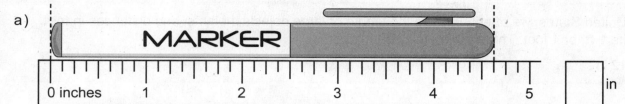 ☐ in

b) 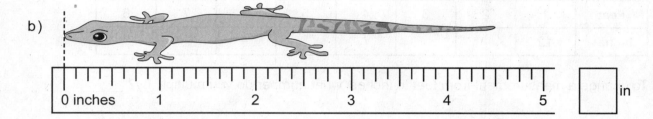 ☐ in

c) 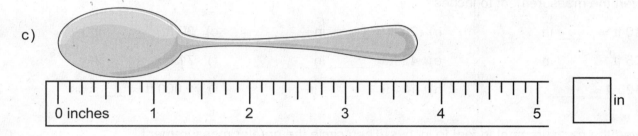 ☐ in

d) 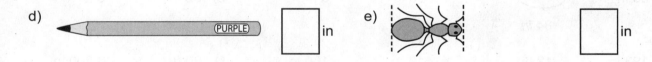 ☐ in   e) ☐ in

f)  ☐ in

g)  ☐ in

7. A penny is about $\frac{1}{16}$ in thick.

   a) What is the height of a stack of 10 pennies? _____

   b) How many pennies are in a stack 2 in high? _____

8. a) Draw a line segment with a length between …

   i) 3 in and 4 in   ii) $4\frac{1}{2}$ in and 5 in   iii) $6\frac{1}{2}$ in and $6\frac{3}{4}$ in

   b) Measure the line segments you drew in part a) to the nearest eighth of an inch.

MD5-11 Feet and Inches

In the United States, we often measure the length of *large* objects, heights, and distances in **feet**. We write **1 ft** for 1 foot. There are 12 inches in 1 foot.

1 ft = 12 in

1. Fill in the measurements in inches.

a)

Feet	1	2	3	4	5	6	7	8
Inches	12							

b) To change a measurement from feet to inches, what number do you multiply by? _____

2. Convert the measurement to inches.

a) 10 ft = _____ in

b) 20 ft = _____ in

c) 30 ft = _____ in

d) 35 ft = _____ in

e) 42 ft = _____ in

f) 71 ft = _____ in

g) 125 ft = _____ in

h) 144 ft = _____ in

i) 1,000 ft = _____ in

3. Convert the measurement in feet to inches. Then circle the greater measurement.

a) 60 in   (7 ft)
                84 in

b) 73 in   6 ft

c) 55 in   5 ft

d) 123 in   12 ft

e) 10 ft   100 in

f) 30 ft   3,000 in

There are 12 inches in 1 foot, so 1 inch = $\frac{1}{12}$ of 1 foot.

---

4. a) Fill in the measurements in feet.

Inches	1	2	3	4	5	6	7	8
Feet	$\frac{1}{12}$	$\frac{2}{12} = \frac{1}{6}$						

b) To change a measurement from inches to feet, what number do you divide by? _____

c) Convert the measurement to feet. Write the answer in lowest terms.

i) 5 in = _____ ft

ii) 9 in = _____ ft

iii) 10 in = _____ ft

iv) 12 in = _____ ft

v) 14 in = _____ ft

vi) 15 in = _____ ft

5. Write the mixed measurement as a measurement in feet. Reduce the fractional part to lowest terms.

a) 3 ft 2 in

2 in = $\boxed{\dfrac{1}{6}}$ ft, so

3 ft 2 in = $\boxed{3\dfrac{1}{6}}$ ft

b) 7 ft 4 in

4 in = $\boxed{}$ ft, so

7 ft 4 in = $\boxed{}$ ft

c) 4 ft 8 in

8 in = $\boxed{}$ ft, so

4 ft 8 in = $\boxed{}$ ft

d) 5 ft 5 in = $\boxed{}$ ft

e) 8 ft 11 in = $\boxed{}$ ft

f) 10 ft 6 in = $\boxed{}$ ft

6. Convert the mixed measurement to a measurement in inches.

a) 3 ft = __36__ in, so

3 ft 5 in = __36__ in + __5__ in

= __41__ in

b) 7 ft = _____ in, so

7 ft 7 in = _____ in + __7__ in

= _____ in

c) 9 ft = _____ in, so

9 ft 1 in = _____ in + _____ in

= _____ in

d) 4 ft = _____ in, so

4 ft 6 in = _____ in + _____ in

= _____ in

e) 2 ft = _____ in, so

2 ft 11 in = _____ in + _____ in

= _____ in

f) 100 ft = _____ in

100 ft 3 in = _____ in + _____ in

= _____ in

7. Write the division as a fraction, then reduce to lowest terms.

a) 27 ÷ 12 = $\boxed{\dfrac{27}{12}}$ = $\boxed{\dfrac{9}{4}}$

b) 46 ÷ 12 = $\boxed{}$ = $\boxed{}$

c) 30 ÷ 12 = $\boxed{}$ = $\boxed{}$

8. Divide by 12 to convert the measurement from inches to feet.

a) 27 in

$27 \div 12 = \dfrac{27}{12} = \dfrac{9}{4}$

$9 \div 4 = 2\ R\ 1$, so $\dfrac{9}{4} = 2\dfrac{1}{4}$

27 in = $\boxed{2\dfrac{1}{4}}$ ft

b) 56 in

56 in = $\boxed{}$ ft

c) 33 in

33 in = $\boxed{}$ ft

d) 45 in

e) 77 in

f) 81 in

In the United States, we measure *walking* distances in **yards**. We write **1 yd** for 1 yard.

1 yard = 3 feet   1 yd = 3 ft

1. Fill in the measurements in feet.

a)

Yards	1	2	3	4	5	6	7	8
Feet	3							

b) To change a measurement from yards to feet, what number do you multiply by? _____

2. Convert the measurements to feet.

a)

yd	ft
10	
20	
30	

b)

yd	ft
22	
35	
81	

c)

yd	ft
220	
1,000	
1,760	

3. Convert the measurement from yards to feet. Write your answer as a proper fraction or as a mixed number.

a) $1\dfrac{1}{2}$ yd $= \dfrac{3}{2}$ yd

$= \dfrac{3}{2} \times 3\ ft = \dfrac{9}{2}\ ft = 4\dfrac{1}{2}\ ft$

b) $\dfrac{1}{4}$ yd

c) $\dfrac{3}{4}$ yd

d) $\dfrac{1}{2}$ yd

e) $5\dfrac{3}{4}$ yd

f) $3\dfrac{1}{8}$ yd

4. Convert the measurement in yards to feet. Then circle the greater measurement.

a) (30 ft)   7 yd

             21 ft

b) 3 yd   6 ft

c) 5 yd   50 ft

d) 12 yd   60 ft

e) 5 ft $1\dfrac{1}{2}$ yd

f) 10 ft $3\dfrac{1}{4}$ yd

5. a) Fill in the measurements in yards.

Feet	1	2	3	4	5	6	7	8	9
Yards	$\frac{1}{3}$								

b) To change a measurement from feet to yards, what number do you divide by? _____

c) Convert the measurement from feet to yards.

i) 11 ft   ii) 17 ft   iii) 22 ft

$$11 \div 3 = \frac{11}{3} \ yd = 3\frac{2}{3} \ yd$$

6. a) Make a conversion table from yards to feet and then to inches. Remember: 1 ft = 12 in.

Yards	1	2	3	4	5	6	7	8
Feet	3							
Inches	36							

b) Use the table to write a measurement in whole inches between …

i) 3 yd and 4 yd _____   ii) 5 yd and 6 yd _____   iii) 7 yd and 8 yd _____

c) Use the table to write a measurement in whole yards between …

i) 105 in and 115 in _____   ii) 160 in and 200 in _____   iii) 200 in and 250 in _____

7. A football field is a rectangle that is 120 yd long and 160 ft wide.

a) What is the length of the field in feet? _____

b) What is the width of the field in yards? _____

c) Find the perimeter of the field (the distance around it) in yards and in feet. Check that you get the same distance both ways.

8. a) A window is 3 ft 2 in wide. How wide is the window in inches?

b) To make a curtain with folds of cloth that cover a window, you need cloth at least three times as wide as the window. What width of cloth do you need?

c) Stores sell cloth in yards and eighths of a yard. What width of cloth should you buy?

d) Cloth costs $15 per yard. How much will you pay for the cloth?

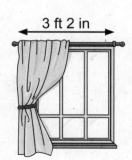

3 ft 2 in

In the United States, we measure *long* distances in **miles.** We write **1 mi** for 1 mile.

1 mile = 1,760 yards   1 mi = 1,760 yd

1. To change a measurement from miles to yards, what number do you

    multiply by? _____

2. Convert the measurement in miles to yards.

    a) 2 mi = _____ yd   b) 4 mi = _____ yd   c) 10 mi = _____ yd

3. Convert the measurement from miles to yards.

    a) $1\frac{1}{2}$ mi = $\frac{3}{2}$ *mi*   b) $\frac{1}{4}$ mi   c) $\frac{3}{4}$ mi

    $$= \frac{3}{2} \times 1,760 \ yd$$

    $$= \frac{5,280}{2} \ yd = 2,640 \ yd$$

    d) $\frac{1}{2}$ mi   e) $\frac{3}{8}$ mi   f) $1\frac{1}{8}$ mi

4. Estimate to circle the greater distance. Then check your estimate by converting
    the measurement in miles to yards.

    a) 3 mi   3,000 yd   b) 50,000 yd   5 mi   c) 3,500 yd   2 mi

    d) 1,000 yd $\frac{5}{8}$ mi   e) 2,800 yd $1\frac{1}{2}$ mi   f) 5,900 yd $3\frac{1}{4}$ mi

5. **a)** Write the measurement as a fraction in lowest terms. Then convert the measurement from miles to yards.

i) 0.6 mi $= \dfrac{6}{10}$ mi $= \dfrac{3}{5}$ mi   ii) 0.1 mi   iii) 0.3 mi

$\qquad\qquad = \dfrac{3}{5} \times 1{,}760$ yd

$\qquad\qquad = \dfrac{5{,}280}{5}$ yd

$\qquad\qquad = 1{,}056$ yd

iv) 0.9 mi   v) 1.2 mi   vi) 1.5 mi

REMINDER ▶ You can change order in multiplication.

Example:

$0.12 \times 1{,}760 = 1{,}760 \times 0.12$
$\qquad\qquad\qquad = 211.2$

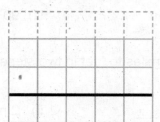

		1	1		
	1	7	6	0	← 0 decimal places
×		0.1	2		← 2 decimal places
	3	5	2	0	
1	7	6	0	0	
2	1	1.2	0		← 0 + 2 = 2 decimal places

---

b) Multiply 1,760 by the decimal to convert the measurement from miles to yards.

i) 0.6 mi $= \underline{\ \ 1{,}056\ \ }$ yd   ii) 0.1 mi $= \underline{\qquad}$ yd   iii) 0.3 mi $= \underline{\qquad}$ yd

		4	3	
	1	7	6	0
×			0.6	
1	0	5	6.0	

iv) 0.9 mi   v) 1.2 mi   vi) 1.5 mi

c) Compare your answers in parts a) and b). Did you get the same answers?

d) Which method do you find easier to convert decimal measurements? Explain.

6. Horse races are measured in *furlongs*. One furlong is $\dfrac{1}{8}$ of a mile.

a) The Kentucky Derby in Louisville, KY, is the most famous horse race in the United States. It is 10 furlongs long. How many yards long is it?

b) The Preakness Stakes race in Baltimore, MD, is $9\dfrac{1}{2}$ furlongs long. How many miles long is that?

c) How many yards long is the Preakness Stakes?

d) How much longer is the Kentucky Derby than the Preakness Stakes?

MD5-14 Changing US Customary Units of Length

REMINDER ▶	1 mi = 1,760 yd	1 yd = 3 ft	1 ft = 12 in

1. a) Convert the measurement from yards to inches in two steps: from yards to feet and then from feet to inches.

    i)   5 yd = _____ ft = _____ in   ii) 10 yd = _____ ft = _____ in

    iii) 25 yd = _____ ft = _____ in   iv) 40 yd = _____ ft = _____ in

b) What number do you multiply by to convert a measurement from yards to inches in *one* step? _____

c) Convert the measurement from yards to inches.

    i)   12 yd = _____ in   ii) 20 yd = _____ in

    iii) 0.5 yd = _____ in   iv) $\frac{1}{2}$ yd = _____ in

d) Which of the two answers in part c) are the same? Why should they be the same?

_____

e) How can you convert a measurement from inches to yards? _____

f) Convert the measurement from inches to yards. Write the answer as a mixed number or a fraction in lowest terms.

    i)   27 in   ii) 45 in   iii) 90 in   iv) 120 in

2. a) Convert the measurement from miles to feet in two steps: from miles to yards and then from yards to feet.

    i)   1 mi = _____ yd = _____ ft   ii) 2 mi = _____ yd = _____ ft

    iii) 10 mi = _____ yd = _____ ft   iv) 0.1 mi = _____ yd = _____ ft

b) What number do you multiply by to convert a measurement from miles to feet in *one* step? _____

c) Convert the measurement from miles to feet.

    i)   5 mi = _____ ft   ii) 1.5 mi = _____ ft

    iii) $\frac{1}{8}$ mi = _____ ft   iv) $\frac{3}{4}$ mi = _____ ft

    v) 1$\frac{1}{2}$ mi = _____ ft   vi) 0.75 mi = _____ ft

d) Which of the answers in part c) are the same? Why should they be the same?

REMINDER ▶

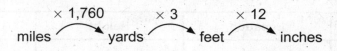

miles $\xrightarrow{\times\ 1,760}$ yards $\xrightarrow{\times\ 3}$ feet $\xrightarrow{\times\ 12}$ inches

3. a) Convert 1 mi to inches.

   1 mi = _____ yd = _____ ft = _____ in

   b) What number do you multiply by to convert a measurement from miles to inches
   in *one step*? _____

   c) Convert the measurement from miles to inches in one step.

   i) 3 mi = _____ in   ii) 10 mi = _____ in   iii) 0.1 mi = _____ in

REMINDER ▶ Perimeter is the distance around a shape.

4. a) Convert the side measurement in yards to feet. Find the perimeter of the
   rectangle in feet.

   i)
   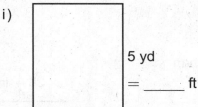
   5 yd
   = _____ ft

   12 ft = _____ yd

   ii)
   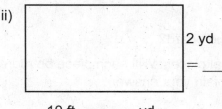
   2 yd
   = _____ ft

   10 ft = _____ yd

   iii)
   $5\frac{1}{2}$ yd
   = _____ ft

   17 ft = _____ yd

   Perimeter = _____ ft   Perimeter = _____ ft   Perimeter = _____ ft

   b) Convert the perimeter from feet to yards.

   i) Perimeter = _____ yd   ii) Perimeter = _____ yd   iii) Perimeter = _____ yd

   c) Convert each side measurement in feet to yards. Find the perimeter of
   each rectangle in yards. Did you get the same answer as in part b)?
   If not, find your mistake.

5. A snail crawled 27 in along a branch and 2 ft along a tree trunk. How far did it crawl
   in total? Write your answer in inches, feet, and yards.

6. An adult polar bear is about 8 times as large as a newborn.

   a) A female newborn polar bear is $12\frac{1}{2}$ in long. How many
   feet long is a female adult polar bear?

   b) A male adult polar bear is 9 ft long. How many inches
   long is a male newborn polar bear?

Measurement and Data 5-14

7. a) Convert the heights of the mountains to feet.

Mountain	Location	Height	Height (ft)
Mount McKinley (Denali)	Alaska	3 mi 4,480 ft	
Mount Whitney	California	2 mi 1,307 yd	
Mauna Kea	Hawaii	2 mi 3,236 ft	

b) How many feet taller is Mount McKinley than the other two mountains?

c) The mountain Aconcagua, in Argentina, is the tallest mountain in South America. It is 2,521 ft taller than Mount McKinley. How tall is Aconcagua in feet?

d) Estimate: Is Aconcagua more than 4 miles tall? Is it more than 5 miles tall? Explain your estimate.

8. A board is 7 ft 4 in long.

a) How many inches long is the board?

b) The board is cut into 4 equal pieces. Will each piece be more than 2 ft long? Estimate and explain your answer.

c) How long is each piece?

9. The bookshelf shown is made from boards that are $\frac{3}{4}$ in thick.

All the openings are the same size. What are the width and the height of each opening in the shelf?

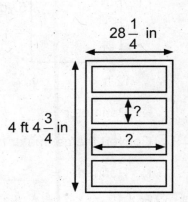

10. Health experts say that a person should walk about 10,000 steps every day. Jose used a step-counting device and found that he walks only 4,700 steps every day.

a) How many more steps does Jose need daily?

b) Jose likes jogging and running. Each step while jogging and running is about 2 ft long. If he jogged the number of steps in part a), how many feet would he jog?

c) Jose sets a routine to add more steps: he jogs to the park a $\frac{1}{2}$ mile away, runs around a path $1\frac{1}{4}$ miles long, and jogs back home. How many miles is that in total?

d) If Jose adds his walking steps and his new routine, does he reach his goal of 10,000 steps per day?

MD5-15 Pounds and Ounces

In the United States, we often measure the mass of *small* objects in **ounces**. We write **1 oz** for 1 ounce.

A slice of bread and a quarter of a small apple each weigh about 1 oz. Five quarters (25¢ coins) weigh about 1 oz.

1. a) What is the mass of 10 quarters? _____ 50 quarters? _____

   b) What is the mass of a whole apple? _____ 3 whole apples? _____

In the United States, we often measure the mass of *large* objects in **pounds**. We write **1 lb** for 1 pound.

There are 16 ounces in 1 pound: 1 lb = 16 oz.

2. a) How many quarters weigh 1 lb? _____   b) How many apples weigh 1 lb? _____

3. a) Fill in the table.

Pounds	1	2	3	4	5	6	7	8
Ounces	16							

   b) To change a measurement from pounds (lb) to ounces (oz), what number do you

   multiply by? _____

4. Change the measurement from pounds to ounces.

   a) 9 lb = _____   b) 12 lb = _____   c) 25 lb = _____

   d) 50 lb = _____   e) 100 lb = _____   f) 200 lb = _____

5. Change the mixed number to an improper fraction. Then convert the measurement to ounces.

   a) $3\frac{1}{4}$ lb $= \frac{13}{4} \times 16\,oz$ b) $1\frac{7}{8}$ lb =   c) $2\frac{1}{2}$ lb =

   $= 13 \times (16\,oz \div 4)$

   $= 13 \times 4\,oz$

   $= 52\,oz$

   d) $1\frac{5}{8}$ lb =   e) $2\frac{1}{16}$ lb =   f) $4\frac{3}{16}$ lb =

6. Convert the measurement in pounds to ounces. Then circle the greater measurement.

a) 30 oz   (2 lb)
            32 oz

b) 33 oz   3 lb

c) 75 oz   5 lb

d) 70 oz   7 lb

e) $3\frac{1}{2}$ lb   50 oz

f) $4\frac{3}{4}$ lb   75 oz

7. a) Fill in the table. Reduce the fractions to lowest terms.

Ounces	1	2	3	4	5	6	7	8
Pounds	$\frac{1}{16}$	$\frac{2}{16}=\frac{1}{8}$						

b) To change a measurement from ounces to pounds, what number do you

divide by? _____

REMINDER ▶ $40 \div 16 = 2 \text{ R } 8$, so $40 \div 16 = \frac{40}{16} = 2\frac{8}{16} = 2\frac{1}{2}$

8. Change the measurement from ounces to pounds. Write your answer as a mixed number in lowest terms.

a) 9 oz =

b) 12 oz =

c) 24 oz =

d) 50 oz =

9. Change the mixed measurement to a measurement in ounces.

a) 3 lb = __48__ oz,

so 3 lb 6 oz = __48__ oz + __6__ oz

= __54__ oz

b) 7 lb = _____ oz,

so 7 lb 7 oz = _____ oz + __7__ oz

= _____ oz

c) 9 lb 4 oz

d) 4 lb 2 oz

e) 2 lb 12 oz

f) 7 lb 8 oz

10. Change the mixed measurement to a measurement in pounds.

a) 6 oz = $\frac{6}{16} = \frac{3}{8}$ lb,

so 3 lb 6 oz = $3\frac{3}{8}$ lb

b) 9 oz =

so 9 lb 9 oz =

c) 7 lb 4 oz

d) 6 lb 12 oz

e) 3 lb 2 oz

f) 5 lb 8 oz

MD5-16 Word Problems with Mass and Length

REMINDER ▶	1 mi = 1,760 yd	1 yd = 3 ft	1 ft = 12 in	1 lb = 16 oz

1. a) A pair of shoes in a box weighs 12 oz. How many pounds will 300 boxes of shoes weigh?

    b) A box of banana bread mix weighs 13.7 oz. How many pounds will 24 boxes weigh?

2. A muffin weighs $3\frac{1}{4}$ oz. A café sold 1,500 muffins. How many pounds of muffins did the café sell?

3. A mail carrier has 250 letters and 2 parcels in a bag. Each letter weighs $\frac{3}{4}$ oz. One parcel weighs $1\frac{3}{4}$ lb and the other weighs $\frac{7}{8}$ lb. What weight is in the mail carrier's bag?

4. In the Adirondacks, a male beaver weighs 45 lb 3 oz. A female beaver weighs 42 lb 8 oz.

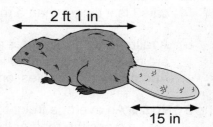

2 ft 1 in

15 in

    a) How much heavier is the male beaver than the female?

    b) How much do the two beavers weigh together?

    c) How long is an average beaver from its nose to the end of its tail?

5. A recipe includes the ingredients below and makes 8 servings of pasta sauce. How much would each serving weigh?

1 lb ground meat	1 lb 6 oz tomato sauce
3 oz onion	4 oz bell peppers
$1\frac{3}{4}$ lb tomatoes	1 oz spices

6. A small cereal box weighs 19 oz and costs $4.00. A large box of the same cereal weighs 1.5 lb and costs $4.50.

    a) Which costs more: 9 small boxes or 8 large boxes of cereal?

    b) Which weighs more: 9 small boxes or 8 large boxes of cereal?

    c) Which cereal box size is cheaper by the ounce?

7. A bag of A-brand muffin mix contains 6.5 oz of mix and costs $1.40. A bag of B-brand muffin mix contains 14 oz of mix and costs $2.80. Which mix is cheaper by the ounce? Explain.

8. A bookshelf is 80 in tall. You can add a 20 in extension on top of the bookshelf. The ceiling is 8 ft high. Should you buy the shelf with or without the extension?

extension→

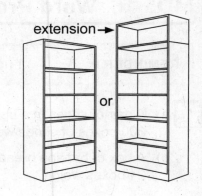

or

9. a) The upper bunk of a bunk bed is $68\frac{5}{8}$ in high. If Mia sits up in bed, she sits 30 in tall. If Mia sits on the upper bunk, what height will her head reach?

   b) The ceiling in Mia's bedroom is 8 ft high. Is the bunk bed a good choice for Mia's room?

   c) Ben is 32 in tall when he sits up in bed. The ceiling in his bedroom is 9 ft high. Is the bunk bed a good choice for Ben's room?

10. a) An average freight train has 70 cars with the same length as the ones in the picture. How long is a row of 70 such cars? Is it longer than a mile?

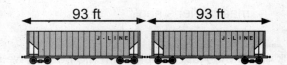

93 ft   93 ft

   b) A railroad company tests a longer train, with 295 cars that are each 62 ft long. How long is the new row of cars? About how many miles long is it?

   BONUS ▶ An average freight train crosses a road in about 7 minutes. All trains cross the road at the same speed. About how many minutes will it take the long train to cross the road?

11. The table lists the lengths and weights of some state animals. Rank the animals by length from smallest to largest. Rank the animals by weight from smallest to largest.

State	Animal	Length	Weight
Kansas	American buffalo (bison)	8 ft 7 in	1,400 lb
Maine	moose	9 ft	1,322 lb
Montana	grizzly bear	$6\frac{5}{6}$ ft	600 lb
Pennsylvania	white-tailed deer	71 in	3,200 oz
Utah	Rocky Mountain elk	7.5 ft	900 lb

Line plots are used to display measurement data. Each line plot has a title, a number line, and a label. Each X on this line plot represents the length of one student's pencil to the nearest inch.

Lengths of Pencils ◄——————— title

		×			
		×			
	×	×	×		
×	×	×	×		
×	×	×	×	×	×

3   4   5   6   7   8   ◄— number line

Length (in) ◄——————— label

1. a) How many students had pencils 4 in long? _____

   b) What was the most common length of pencil? _____

   c) How many pencils were measured? _____

   d) If all the pencils were laid end to end, how long would the line be if measured ...

   i) in inches? _____

   ii) in feet? _____

2. A teacher measured student heights to the nearest inch. Create a line plot to display the data below. Include a title, a number line, and a label.

53	54	57	56	57	56	53	56	52	61
55	53	54	53	55	55	57	60	53	61

Title: _____

52   53

Label: _____

a) What was the most common height in inches? _____

b) What was the tallest height measured in feet and inches? _____

3. The line plot shows the number of golf courses of different lengths when measured to the nearest mile.

Lengths of Golf Courses

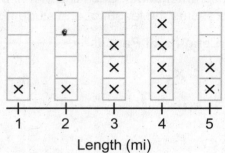

Length (mi)

a) What is the most common length for a golf course? _____

b) What is the length of the longest course in yards? _____

c) A golfer walks twice the length of the course during a round of golf.

How many yards does the golfer walk on a 3 mi course? _____

4. A class weighed their backpacks to the nearest pound. Use the data below to create a line plot. Include the title, number line, and label.

14	14	18	17	16	17	15	14	16	19
14	15	17	15	16	15	14	19	17	18

Title: _____

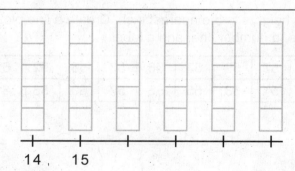

14 , 15

Label: _____

a) Remember: 1 lb = 16 oz. What was the heaviest backpack weight in ounces? _____

b) How many students weighed their backpacks? _____

c) What was the total weight of the backpacks? _____

d) Madison's backpack weighs 244 oz. Convert to pounds and add it to the line plot. _____

e) Doctors recommend the maximum weight of a backpack should be about 15 lb. What do you recommend for this class? _____

MD5-18 Line Plots with Fractions

1. Students measured the distance they each walk to school. The line plot shows the data rounded to the nearest eighth of a mile.

Distance Walked to School

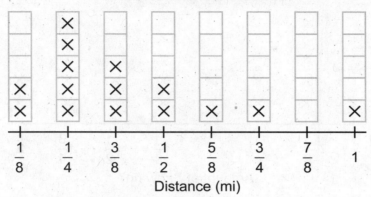

a) How many students measured the distance? _____

b) How many students walked more than $\frac{1}{4}$ mi to school? _____

c) Remember: 1 mile = 1,760 yards. For the students who walked $\frac{3}{8}$ mi, how many yards did each walk? _____

d) What was the total distance in miles walked by all the students? _____

2. A hospital weighed newborn babies to the nearest eighth of a pound.

$7\frac{1}{8}$	$7\frac{1}{4}$	8	$7\frac{1}{8}$	$8\frac{1}{4}$	$7\frac{1}{2}$	$7\frac{3}{4}$	$7\frac{1}{2}$	$7\frac{1}{2}$	8
$7\frac{1}{2}$	$7\frac{3}{8}$	$7\frac{1}{4}$	8	$8\frac{1}{4}$	$7\frac{3}{8}$	$7\frac{1}{4}$	$7\frac{3}{4}$	$7\frac{1}{2}$	$8\frac{1}{8}$

Complete the line plot to show the data and then answer the questions:

Weights of Newborns

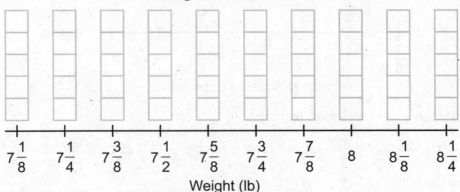

a) What was the most common baby weight? _____

b) How many babies weighed more than 8 lb? _____

c) What is the difference in weight between the lightest and heaviest babies? _____

d) How many ounces did the lightest baby weigh? _____

3. The line plot shows the daily snowfall to the nearest quarter of an inch during April in Valdez, AK.

Daily Snowfall in April

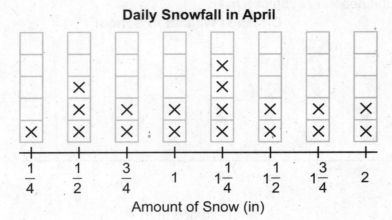

Amount of Snow (in)

a) The month of April has 30 days. How many days had less than $\frac{1}{4}$ in of snow? _____

b) Gail says $20\frac{3}{4}$ in of snow fell in April. Is she correct?

c) If the monthly snowfall for each month from September to May is $20\frac{3}{4}$ in, what would the total annual snowfall be? _____

4. A Grade 5 class ran laps to raise money for a charity. The line plot shows the number of laps run by each student.

Running Laps for Charity

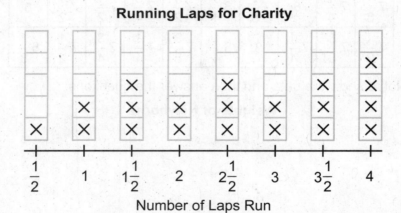

Number of Laps Run

a) How many students ran more than two laps? _____

b) One lap is $\frac{1}{4}$ mi. Multiply each number of laps run by $\frac{1}{4}$ to find the different distances run by students: $\frac{1}{8}$ mi, $\frac{1}{4}$ mi, $\frac{3}{8}$ mi, _____

c) How many students ran less than 1 mi? _____

d) What is the total number of laps that all students ran? _____

e) What is the total distance in miles that all students ran? _____

MD5-19 Converting Time

1 hour (h) = 60 minutes (min)	1 minute (min) = 60 seconds (s)

1. Find the number of minutes.

a) 2 h 36 min   b) 3 h 14 min   c) 10 h 32 min

    = (2 × 60 + 36) min

    = (120 + 36) min

    = 156 min

2. Convert seconds to minutes by dividing by 60.

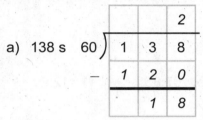 a) 138 s

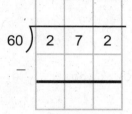 b) 272 s

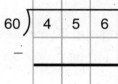 c) 456 s

    __2__ min   __18__ s   ____ min ____ s   ____ min ____ s

A 24-hour clock has two scales for hours.

The outer scale is for times from midnight until just before noon (a.m.).
The inner scale is for times from noon until just before midnight (p.m.).

We always write two digits for the hour. We do not write a.m. or p.m.

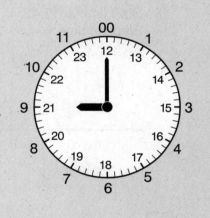

To convert to 24 h format, add 12 to the hours for times from 1:00 p.m. to 11:59 p.m.

To convert to 12 h format, subtract 12 from the hours for times from 13:00 to 23:59.

Example: 2:15 p.m. = 14:15   Example: 19:25 = 7:25 p.m.

---

3. Convert to 24 h format.

a) 4:15 p.m.   b) 7:20 p.m.   c) 11:54 p.m.   d) 9:45 a.m.   e) 11:54 a.m.   f) 12:25 a.m.

  _____   _____   _____   _____   _____   _____

4. Write the time using a.m. or p.m.

a) 16:30   b) 14:48   c) 23:15   d) 3:35   e) 12:15   f) 00:58

  _____   _____   _____   _____   _____   _____

Milena started her homework at 14:30. She worked
for 1 hour 10 minutes (or 1:10). This is called the
elapsed time. When did she finish?

	1	4	:	3	0	← start time
+		1	:	1	0	← elapsed time
	1	5	:	4	0	← end time

5. Find the end time.

a) Start time = 12:10
   Elapsed time = 4:25

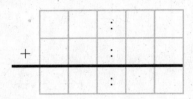

b) Start time = 8:40
   Elapsed time = 3:15

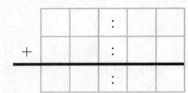

c) Start time = 9:12
   Elapsed time = 4:37

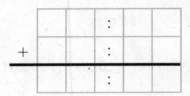

6. Regroup 60 minutes as one hour.

a) 16:95   b) 09:72   c) 13:86   d) 14:97   e) 07:85   f) 12:99

   _17:35_   _____   _____   _____   _____   _____

7. a) Find the end time. Regroup where necessary.

i) Start time = 12:43
   Elapsed time = 4:38

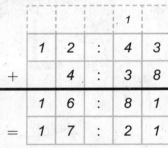

ii) Start time = 9:36
   Elapsed time = 3:53

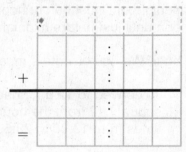

iii) Start time = 11:13
   Elapsed time = 1:58

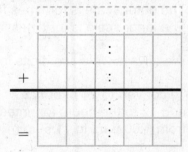

iv) Start time = 11:45
    Elapsed time = 3:30

v) Start time = 9:35
   Elapsed time = 5:31

vi) Start time = 10:13
    Elapsed time = 6:27

b) Rewrite the end time in Question 7.a) using a.m. or p.m.

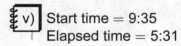

   i) _5:21 p.m._ ii) _____ iii) _____ iv) _____ v) _____ vi) _____

8. A hockey game that started at 11:35 a.m. lasted for 2 hours 15 minutes.
What time did it end? Rewrite the time using a.m. or p.m.

MD5-20 Calculating Elapsed Time

To find the **elapsed time** from 08:10 to 14:25,
we can subtract. 6 hours 15 minutes has elapsed.

	1	4	:	2	5	← end time
−	0	8	:	1	0	← start time
		6	:	1	5	← elapsed time

1. Find the elapsed time.

a) 09:12 to 14:18

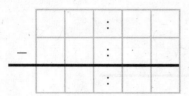

b) 05:34 to 18:49

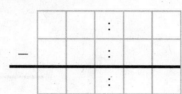

c) 08:35 to 19:48

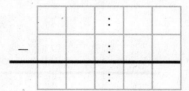

d) 04:18 to 16:44

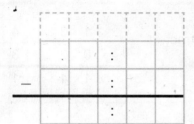

e) 06:28 to 09:53

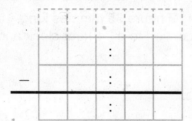

f) 12:47 to 14:52

2. Regroup 1 hour as 60 minutes.

a) 4:25

___3:85___

b) 3:13

_____

c) 8:08

_____

d) 14:25

_____

e) 13:13

_____

f) 18:39

_____

Example: Find the elapsed time from 07:35 to 14:15.
Sam couldn't subtract 35 min from 15 min, so he regrouped 1 hour as 60 minutes.

Step 1: Regroup 1 hour as 60 minutes.

14:15 ⟶ 13:75

Step 2: Subtract

	1	3	:	7	5
−	0	7	:	3	5
		6	:	4	0

3. Find the elapsed time by regrouping first.

a) 05:35 to 13:24

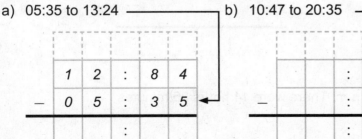

b) 10:47 to 20:35

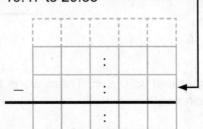

c) 8:49 to 14:16

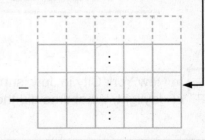

4. Find the elapsed time by first converting to a 24-hour clock. Regroup where necessary.

a) 8:30 a.m. to 2:55 p.m.

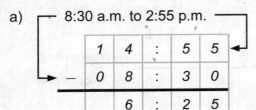

	1	4	:	5	5
−	0	8	:	3	0
		6	:	2	5

b) 7:25 a.m. to 3:45 p.m.

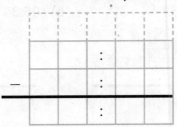

c) 9:28 a.m. to 1:55 p.m.

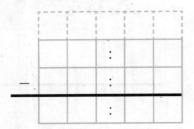

d) 10:45 a.m. to 5:33 p.m.

A movie ended at 1:45 p.m. and was 2 hours 35 minutes long.
To find the start time of the movie, convert to a 24-hour clock and subtract.

The movie started at 11:10 a.m.

	1	3	:	4	5
−	0	2	:	3	5
	1	1	:	1	0

5. Find the start time. Regroup where necessary.

a) Elapsed time = 3:47
   End time = 11:36 a.m.

	1	1	:	3	6
−		3	:	4	7
			:		

b) Elapsed time = 8:27
   End time = 11:15 p.m.

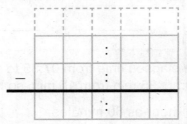

c) Elapsed time = 0:38
   End time = 1:25 p.m.

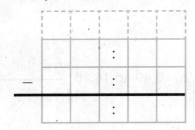

d) Elapsed time = 12:15
   End time = 8:10 p.m.

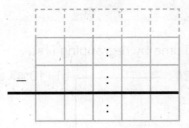

6. In New York City in July, sunset was at 8:20 p.m. There were 14 hours 35 minutes of daylight. What time did the sun rise?

MD5-21 Time Word Problems

1. Ethan mowed the lawn starting at 9:40 am. He worked for 2 hours 15 minutes. What time did he finish?

2. A baseball game started at 11:25 a.m. and lasted for 2 hours 45 minutes. What time did it end?

3. Keesha did homework from 3:25 p.m. to 5:43 p.m. For how long was she doing homework?

4. Grace's family drove from Fort Worth, TX, to Houston. They left at 10:05 a.m. and arrived at 1:58 p.m. For how long did they travel?

5. A bicycle rental company charges per hour for renting a bike. Jayden rents a bike starting at 9:45 a.m. and returns it at 3:45 p.m.

   a) For how long did he have the bike?

   b) The rental company charges $5 plus $10 per hour. What is the total charge?

6. Nao fills out a timesheet at work. It shows when he starts and ends work.

Day	Start time	End time	Time Worked	
Monday	8:25 a.m.	11:40 a.m.	_____ h	_____ min
Tuesday	12:15 p.m.	2:45 p.m.	_____ h	_____ min
Wednesday	1:45 p.m.	5:15 p.m.	_____ h	_____ min
Thursday	9:45 a.m.	5:10 p.m.	_____ h	_____ min
Friday	8:25 a.m.	4:00 p.m.	_____ h	_____ min

   a) Find the amount of time he worked each day.

   b) What is the total time he worked for the week written in hours and minutes?

   c) Write the total number of hours Nao worked as an improper fraction.

   d) Nao earns $12 per hour. Calculate his week's pay by multiplying the hourly wage by the total number of hours written as a fraction.

   e) Write the total number of hours worked as a decimal.

   f) Calculate his week's pay by multiplying the hourly wage by the total number of hours written as a decimal.

   g) Are the answers in part d) and part f) the same? If not, find your mistake.

MD5-22 Diagrams and Fractions

Sally had $31. She spent $\frac{2}{5}$ of her money on a book. How much money is left?

Sally draws a diagram with equal blocks. She finds the size of one block: $31.00 \div 5 = 6.20$.

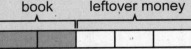

book   leftover money

The leftover money is $3 \times 6.20 = 18.60$.

1. Finish the diagram. Find the size of each block. Then solve the problem.

   a) Jay had $30. He spent $\frac{1}{4}$ of his money on lunch. How much money is left?

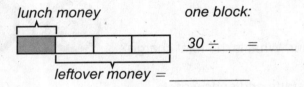

   lunch money   one block:

   $30 \div \underline{\hspace{1cm}} = \underline{\hspace{1cm}}$

   leftover money = _____

   b) Ava had $28. She spent $\frac{3}{7}$ of her money on a gift. How much money is left?

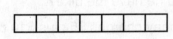

   c) Leo spent $4.50 on a snack. This was $\frac{1}{4}$ of his money. How much money did he have in the beginning?

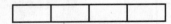

   d) Kim spent $\frac{2}{3}$ of her money on a music album. The album cost $10.50. How much money did she have in the beginning?

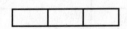

   e) Tom spent $6.75 on lunch. This was $\frac{3}{7}$ of his money. How much money does he have left?

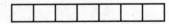

   f) Yu spent $\frac{2}{7}$ of her money on a fruit smoothie. She has $12.50 left. How much money did she have in the beginning?

2. Alice bought paint for $8.30, a canvas for $2, and two paintbrushes for 99¢ each.

   a) How much money did she spend? _____

   b) She spent $\frac{2}{5}$ of her money on art supplies. How much money did she have in the beginning?

Ron spent $\frac{1}{8}$ of his money on lunch and $\frac{3}{7}$ of the leftover money on a book. He drew a diagram to show his spending.

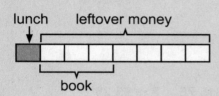

The **total** money is divided into 8 parts.

3. Label the parts in the diagram.

   a) Ravi spent $\frac{3}{8}$ of his break time eating lunch and $\frac{2}{5}$ of the rest of the break helping in the library.

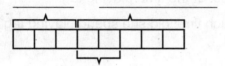

   b) Nina used $\frac{3}{7}$ of a pack of rice to make curry and $\frac{1}{4}$ of the rest to make rice pudding.

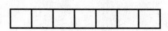

   c) Ben donated $\frac{3}{5}$ of his money to charity and spent half the rest on a new phone.

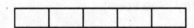

   d) Maria spent $\frac{3}{7}$ of her free time reading and half the rest of the time playing ball.

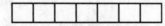

4. In Question 3, circle the fraction that tells you how many blocks should be in the diagram.

5. Draw a diagram for each situation. Label the parts of the diagram.

   a) John spent $\frac{1}{4}$ of the summer volunteering. He spent $\frac{2}{3}$ of the rest of his summer at camp.

   b) Grace used $\frac{1}{6}$ of a bag of flour to bake a cake. She used $\frac{3}{5}$ of the rest to make cookies.

Ron spent $\frac{1}{8}$ of his money on lunch and $\frac{3}{7}$ of the leftover money on a book. The book cost $15.75. How much money did he have in the beginning?

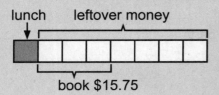

lunch   leftover money

book $15.75

Ron finds the size of one block.

$15.75 = 3 blocks, so 1 block = $15.75 ÷ 3 = $5.25

8 blocks = 8 × $5.25 = $42.00. Ron had $42 in the beginning.

6. Draw the diagram and label its parts. Find the size of one block. Then solve the problem.

a) Josh had $15. He spent $\frac{1}{4}$ of his money on a snack. He spent $\frac{2}{3}$ of the rest on a new scarf. How much money does he have left?

one block: _____

Josh has _____ left.

b) Ava spent 1 hour 10 minutes outside. She spent $\frac{3}{7}$ of the time playing soccer and $\frac{3}{4}$ of the rest of the time playing tag.

How much time did she spend playing tag?

one block: _____

Ava spent _____ minutes playing tag.

c) Sofia bought a carton of juice for $3.50, which was $\frac{1}{5}$ of her money. She spent half the rest of her money on fruit. How much money did she have left?

d) Jayden spent $\frac{2}{5}$ of his money on an album and a poster. The album cost $8.40, and the poster cost $3. He spent $\frac{1}{3}$ of the rest of his money on a book. How much money did he have in the beginning? How much did the book cost?

e) Sam had 3 lb of flour. He used $\frac{3}{8}$ of it to make dumplings, and half the rest to bake a loaf of bread. How many ounces of flour are left?

f) Anika spent $\frac{1}{7}$ of her vacation at a cottage. She spent $\frac{2}{3}$ of the rest of the vacation volunteering. She spent the remaining 10 days at camp. How many weeks long was her vacation?

1. Write an addition equation and a multiplication equation for the number of squares in the rectangle.

a)

_____2 + 2 + 2 = 6_____

_____3 × 2 = 6_____

b)

_____

_____

c)

_____

_____

2. How many squares fit along the length and the width of the rectangle? Write a multiplication equation for the number of boxes in the rectangle.

a) Width = _____3_____

Length = _____4_____

_____3 × 4 = 12_____

b) 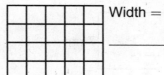 Width = _____

Length = _____

_____

c) 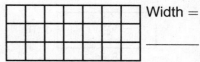 Width = _____

Length = _____

_____

A square with sides 1 cm long is called a **square centimeter** (cm²). 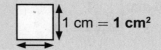 1 cm = **1 cm²**

3. Use a ruler to join the marks and divide the rectangle into square centimeters (cm²). Write a multiplication equation for the number of square centimeters needed to cover the rectangle.

a)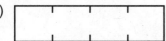

_____ cm²

b)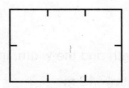

_____ cm²

c)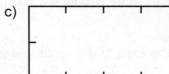

_____ cm²

The **area** of a flat shape is the number of squares of the same size needed to cover the shape without gaps or overlaps. A **square centimeter** (cm²) is a unit for measuring area.

4. Area is also measured in other square units. Predict the names of the units.

a)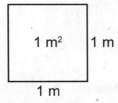
1 m² | 1 m
1 m

_____square meter_____

b)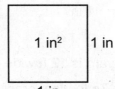
1 ft² | 1 ft
1 ft

_____

c)
1 in² | 1 in
1 in

_____

5. How many unit squares are needed to cover the rectangle? Write a multiplication equation for the area. Include the units.

a)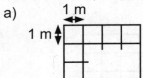

Area = ___4 × 3 = 12 m²___

b)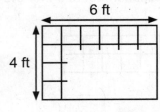

Area = _____

c)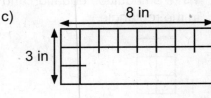

Area = _____

Area of a rectangle = length × width

6. a) Calculate the area of the rectangle. Include the units.

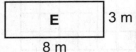

| E | 3 m |
8 m

Area = _____

| P | 7 cm |
8 cm

Area = _____

6 cm
| O | 6 cm |

Area = _____

6 mm | T |
8 mm

Area = _____

5 km
6 km | A |

Area = _____

| K | 5 m |
8 m

Area = _____

b) Use the letters to list the rectangles from least area to greatest area: ___, ___, ___, ___, ___, ___

What state capital did you get? _____

7. Find the area of the rectangle using the length and the width. Include the units.

a) Length = 9 m
Width = 7 m
Area = ___9 m___ × ___7 m___
= ___63 m²___

b) Length = 12 m
Width = 9 m
Area = _____ × _____
= _____

c) Length = 16 cm
Width = 8 cm
Area = _____ × _____
= _____

d) Length = 27 in
Width = 11 in

e) Length = 39 ft
Width = 12 ft

f) Length = 33 yd
Width = 12 yd

8. a) A backyard is 12 ft wide and 25 ft long. What is the area of the backyard?

b) It costs 7¢ to plant grass seed for 5 ft² of lawn. How much will it cost to plant seed for the whole backyard?

MD5-24 Area of Rectangles with Fractions

1. How many small squares cover the large square? Find the area of each small square.

a)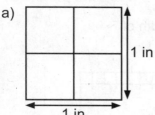

1 in
1 in

_____ small squares

Area = 1 in² ÷ _____ = [ ] in²

b)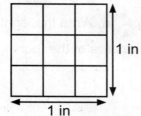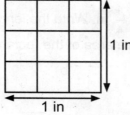

1 in
1 in

_____ small squares

Area = 1 in² ÷ _____ = [ ] in²

c)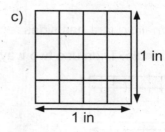

1 in
1 in

_____ small squares

Area = 1 in² ÷ _____ = [ ] in²

2. Find the length and the width of each **small** square. Then find the area of each small square.

a)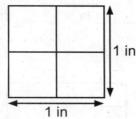

1 in
1 in

Length = $\frac{1}{2}$ in

Width = $\frac{1}{2}$ in

Area = $\frac{1}{2}$ in × $\frac{1}{2}$ in = $\frac{1}{4}$ in²

b)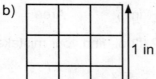

1 in
1 in

Length =

Width =

Area =   ×   =

c)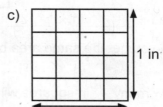

1 in
1 in

Length =

Width =

Area =

3. Compare your answers in Questions 1 and 2. Why should you get the same answer? Explain.

4. a) How many small squares cover the large square? _____

b) Use two ways to find the area of a small square.

   i) Area = 1 m² ÷ _____ =

   ii) Length =   Width =

      Area =   ×   =

c) Did you get the same area both ways? If not, find your mistake.

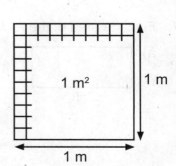

1 m²
1 m
1 m

A square with sides measuring $\frac{1}{4}$ inch has this area: $\frac{1}{4}$ in $\times$ $\frac{1}{4}$ in $=$ $\frac{1}{16}$ in².

5. Each small square has sides representing $\frac{1}{4}$ in. Write the length and the width of the rectangle. Then use two ways to find the area of the rectangle.

a)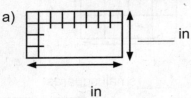

_____ in

_____ in

Area = _____ squares

= _____ $\times$ $\frac{1}{16}$ in² = ☐ in² = _____ in²

Area = _____ in $\times$ _____ in = _____ in²

b)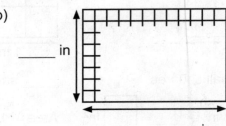

_____ in

_____ in

Area = _____ squares

= _____ $\times$ $\frac{1}{16}$ in² = ☐ in² = _____ in²

Area = _____ in $\times$ _____ in = _____ in²

c) Did you get the same area both ways? If not, find your mistake.

6. a) How many $\frac{1}{4}$ in squares will fit along each side of the rectangle?

Length = _____ squares   Width = _____ squares

b) How many $\frac{1}{4}$ in squares are needed to cover the rectangle? _____

Find the area of the rectangle using the area of one square.

Area = _____ $\times$ ☐ in² = ☐ in²

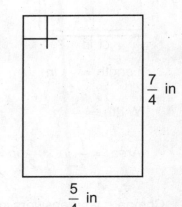

$\frac{7}{4}$ in

$\frac{5}{4}$ in

c) Area of rectangle = length $\times$ width

= ☐ in $\times$ ☐ in = ☐ in²

d) Did you get the same answer in parts b) and c)? If not, find your mistake.

7. Find the area of the rectangle.

a)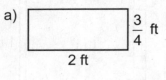

$\frac{3}{4}$ ft

2 ft

Area = _____

= _____

b)

$\frac{1}{2}$ in

$\frac{3}{8}$ in

Area = _____

= _____

c) $1\frac{1}{4}$ mi = $\frac{}{4}$ mi

$3\frac{1}{2}$ mi = $\frac{}{2}$ mi

Area = _____

= _____

Area of rectangle = width × length = length × width   OR   Area = $w \times \ell = \ell \times w$

1. Find the area of the rectangle.

   a) Width = 3 m

      Length = 6 m

      Area = _____

      = _____

   b) Width = 2 km

      Length = 9 km

      Area = _____

      = _____

   c) Width = 6.5 cm

      Length = 8 cm

      Area = _____

      = _____

   d) Width = 3 km

      Length = 6.6 km

      Area = _____

      = _____

   e) Width = 4.2 mm

      Length = 9.3 mm

      Area = _____

      = _____

   f) Width = 2.45 m

      Length = 14 m

      Area = _____

      = _____

2. Write an equation for the area of the rectangle. Then find the unknown width.

   a) Width = w in

      Length = 5 in

      Area = 15 in²

      $w \times 5 = 15$

      $w = 15 \div 5$

      $w = 3$ in

   b) Width = w ft

      Length = 2 ft

      Area = 12 ft²

   c) Width = w mi

      Length = 6 mi

      Area = 24 mi²

3. Write an equation for the area of the rectangle. Then find the unknown length.

   a) Width = 5 cm

      Length = ℓ cm

      Area = 30.5 cm²

   b) Width = 7 km

      Length = ℓ km

      Area = 43.47 km²

   c) Width = 10 m

      Length = ℓ m

      Area = 167.8 m²

4. a) A rectangle has an area of $9\frac{3}{8}$ ft² and a width of 3 ft. What is its length?

   b) A rectangle has an area of 5 cm² and a length of 8 cm. What is its width?

   c) A square has an area of 16 cm². What is its width?

5. Draw a line to divide the shape into two rectangles. Use the areas of the rectangles to find the total area of the shape.

a)

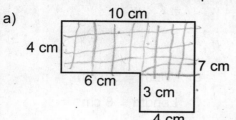

Area of rectangle 1 = __12 cm²__

Area of rectangle 2 = __40 cm²__

Total area = __52 cm²__

b)

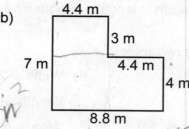

Area of rectangle 1 = __13.2 m²__

Area of rectangle 2 = __35.2 m²__

Total area = __48.4 m²__

c)

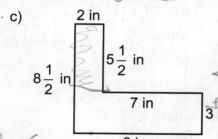

Area of rectangle 1 = __11 in²__

Area of rectangle 2 = __27 in²__

Total area = __38__

d)

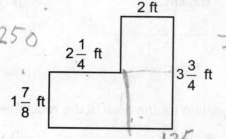

Area of rectangle 1 = __22__

Area of rectangle 2 = __3 3/4__

Total area = _____

REMINDER ▶ Perimeter is the distance around a shape.

---

6. Find the length and the area of the rectangle.

a) Width = 2 cm   Perimeter = 12 cm

Length = __4 cm__

Area = __8 cm²__

b) Width = 4 cm   Perimeter = 20 cm

Length = __6 cm__

Area = __24 cm²__

7. Yu wants to build a rectangular vegetable garden 2 ft wide and with a perimeter of 16 ft.

a) Sketch the garden on the grid. Each square on the grid represents one square foot.

b) What is the length of the vegetable garden?   6 ft

c) Yu wants to build a fence around the garden. Fencing costs $2.50 per foot. How much will the fencing cost? $40.00

d) Yu will plant 8 cucumber seeds on each square foot of land. A pack of 50 seeds costs 99¢. How many packs of seeds does she need?

e) How much will the seeds and fencing cost altogether?

Measurement and Data 5-25

MD5-26 Stacking Blocks

1. How many blocks are in the shaded row?

a)

_3_ blocks

b)

_3_ blocks

c)

_3_ blocks

2. How many blocks are in the shaded row?

a)

_6_ blocks

b)

_6_ blocks

c)

_6_ blocks

3. Write the number of shaded blocks. Then write an addition equation and a multiplication equation for all the blocks.

a)

_4_ blocks shaded

_4_ + _4_ + _4_

= _12_ blocks

_4_ × 3 = _12_ blocks

b)

_12_ blocks shaded

_12_ + _12_ + _12_

= _36_ blocks

_12_ × _3_ = _36_ blocks

c)

_8_ blocks shaded

_8_ + _8_ + _8_ + _8_

= _32_ blocks

_8_ × _4_ = _32_ blocks

4. a) Write a multiplication equation for the number of blocks in one layer.

_3_ × _2_ = _6_ blocks

b) Calculate the number of blocks in the shaded layer. Then calculate the total number of blocks in the stack.

i)

ii)

iii)

blocks in each layer   number of layers

_2_ × _3_ × _2_

= _12_ blocks

_2_ × _3_ × _3_

= _18_ blocks

_2_ × _3_ × _4_

= _24_ blocks

5. Write a multiplication statement for the number of blocks in the stack.

a)

$\underline{2} \times \underline{4} \times \underline{5}$
$= \underline{40}$ blocks

b)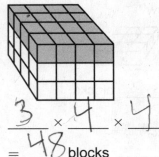

$\underline{3} \times \underline{4} \times \underline{4}$
$= \underline{48}$ blocks

c)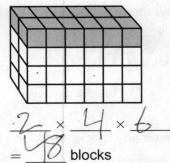

$\underline{2} \times \underline{4} \times \underline{6}$
$= \underline{48}$ blocks

6. a)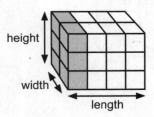

height
width
length

Number of blocks in a vertical layer

= height × width

= $\underline{3}$ × $\underline{2}$ = $\underline{6}$ blocks

Total number of blocks

= height × width × length
= $\underline{3}$ × $\underline{2}$ × $\underline{4}$ = $\underline{24}$ blocks

b)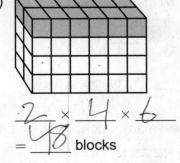

Number of blocks in a vertical layer

= height × width

= $\underline{2}$ × $\underline{3}$ = $\underline{6}$ blocks

Total number of blocks

= height × width × length
= $\underline{2}$ × $\underline{3}$ × $\underline{5}$ = $\underline{30}$ blocks

c)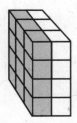

Number of blocks in a vertical layer

= height × width

= $\underline{4}$ × $\underline{3}$ = $\underline{12}$ blocks

Total number of blocks

= height × width × length
= $\underline{4}$ × $\underline{3}$ × $\underline{2}$ = $\underline{24}$ blocks

d)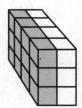

Number of blocks in a vertical layer

= height × width

= $\underline{3}$ × $\underline{4}$ = $\underline{12}$ blocks

Total number of blocks

= height × width × length
= $\underline{3}$ × $\underline{4}$ × $\underline{2}$ = $\underline{24}$ blocks

7. Two stacks in Question 6 have the same number of blocks. Use the height, width, length, and the properties of multiplication to explain why this happens.

MD5-27 Volume

Volume is the amount of space taken up by a three-dimensional (or 3-D) object.

These objects have a volume of 4 cubes.

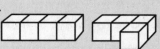

1. Count the number of cubes to find the volume of the object.

a)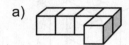

Volume = ___5___ cubes

b)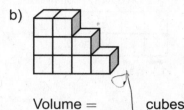

Volume = ___9___ cubes

c)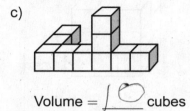

Volume = ___10___ cubes

We measure volume in cubic units or unit cubes. Note: The cubes below are not drawn to scale.

| 1 cm³ = 1 cubic centimeter | 1 m³ = 1 cubic meter | 1 in³ = 1 cubic inch |

  1 cm 1 cm 1 cm

  1 m 1 m 1 m

1 in 1 in 1 in

2. Find the volume of the object made from unit cubes. Include units in your answer.

a) 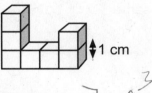 $\updownarrow$ 1 cm

Volume = ___7 cm³___

b) 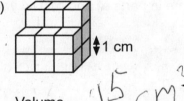 $\updownarrow$ 1 cm

Volume = ___15 cm³___

c) 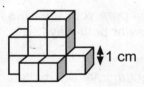 $\updownarrow$ 1 cm

Volume = ___11 cm³___

d) 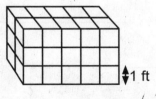 $\updownarrow$ 1 ft

Volume = ___30 ft³___

e) $\leftrightarrow$ 1 in

Volume = ___12 in³___

f) 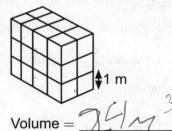 $\updownarrow$ 1 m

Volume = ___24 m³___

g)  4 m

Volume = ___36 m³___

h) 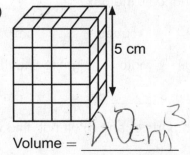 5 cm

Volume = ___20 cm³___

i) 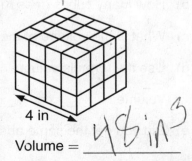 4 in

Volume = ___48 in³___

Mathematicians call rectangular boxes **rectangular prisms**.

3. Use two ways to find the volume of the rectangular prism made from unit cubes.

   a) Find the number of unit cubes. Include units in the answer.

   i)   ii)   iii)

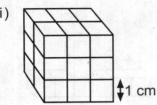 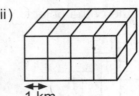 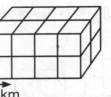 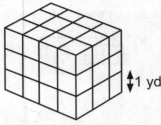

1 cm   1 km   1 yd

Volume = _____   Volume = _____   Volume = _____

   b) Find the length, width, and height of the prisms in part a). Multiply length × width × height to find the volume. Include the units!

   i) Length = _____   ii) Length = _____   iii) Length = _____

       Width = _____   Width = _____   Width = _____

       Height = _____   Height = _____   Height = _____

       Volume = _____   Volume = _____   Volume = _____

   c) Compare your answers for volume in parts a) and b). Did you get the same answer both ways?

For a rectangular prism, volume = length × width × height OR $V = \ell \times w \times h$

4. Kim has a box that is 15 cm long, 10 cm wide, and 8 cm tall.
   She packs the box with 1 cm cubes.

   a) How many cubes fit along each side of the box?

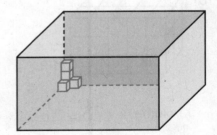

     Length = _____ cubes

     Width = _____ cubes

     Height = _____ cubes

   b) How many cubes does Kim need to fill the box? _____

   c) What is the volume of the box? _____ cubes = _____ cm³

   d) Use the formula volume = length × width × height to find the volume of the box.

     Volume = _____ cm × _____ cm × _____ cm = _____ cm³

   e) Did you get the same answer in parts c) and d)? If not, find your mistake.

5. Find the volume of the prism.

a) Length = ___3 m___

   Width = ___2 m___

   Height = ___2 m___

   Volume = _3 m × 2 m × 2 m_ = ___12 m³___

(prism labeled 2 m, 2 m, 3 m)

b) Length = _____

   Width = _____

   Height = _____

   Volume = _____ = _____

(prism labeled 2 in, 9 in, 4 in)

c) Length = _____

   Width = _____

   Height = _____

   Volume = _____ = _____

(prism labeled 2 ft, 4 ft, 5 ft)

d) Length = _____

   Width = _____

   Height = _____

   Volume = _____ = _____

(prism labeled 9 cm, 5 cm, 5 cm)

e) ℓ = _____

   w = _____

   h = _____

   V = _____ = _____

(prism labeled 6 mm, 6 mm, 11 mm)

f) ℓ = _____

   w = _____

   h = _____

   V = _____ = _____

(prism labeled 4 in, 4 in, 3 in)

6. Find the volume of the rectangular prism.

a) Length 25 m, width 6 m, height 6 m

   Volume = _____ = _____

b) Length 15 ft, width 30 ft, height 45 ft

   Volume = _____ = _____

c) Length 90 cm, width 15 cm, height 8 cm

   Volume = _____ = _____

d) Length 115 in, width 20 in, height 30 in

   Volume = _____ = _____

7. Estimate the answer. Then use a calculator to find the actual value.

a) The tower of the Aon Center in Chicago, IL, is a rectangular prism that is 194 ft wide, 194 ft long, and 1,123 ft tall. What is the volume of the tower?

b) The Cheung Kong Center Tower in Hong Kong, China, is a rectangular prism 154 ft wide, 154 ft long, and 928 ft tall. What is the volume of the tower?

c) Which tower has a greater volume, the Aon Center or the Cheung Kong Center Tower? What is the difference between them?

8. Use this prism to explain why 2 × 3 × 4 = 3 × 2 × 4.

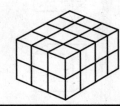

Flat surfaces on a three-dimensional (or 3-D) shape are called **faces**. Faces meet at **edges**. Edges meet at **vertices**. You can show hidden edges with dashed lines.

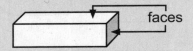

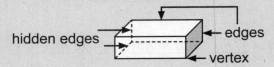

1. Draw dashed lines to show the hidden edges. The dot marks a hidden vertex.

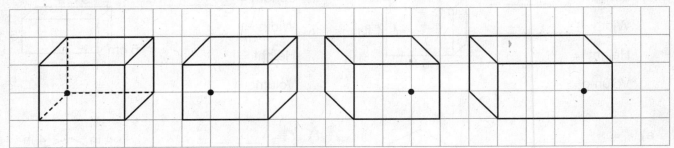

Edges and vertices of a shape make its **skeleton**. This is the skeleton of a cube.

2. Imagine the skeleton covered in paper and placed on a table. Shade the edges that would be hidden.

a)   b)   c) 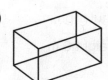   d)

3. Shade the face or faces named below.

a) front face

   i)   ii)   iii)   iv)

b) back face

   i)   ii)   iii)   iv)

c) side faces

   i)   ii)   iii)   iv)

d) top and bottom faces

   i)   ii)   iii)   iv)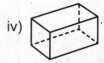

A **net** of a three-dimensional (or 3-D) shape is a pattern that you can fold to make the shape.

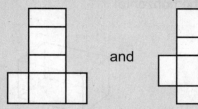 and 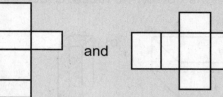 and 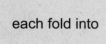 each fold into

4. Draw a net of the prism on the grid below. Label each face. Each square on the grid represents 1 cm.

a)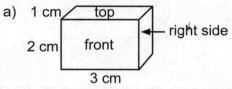

b)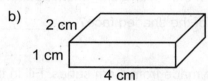

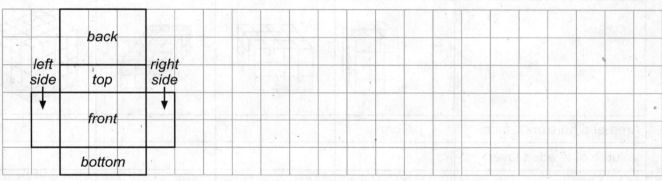

5. On the net, mark the prism that matches.

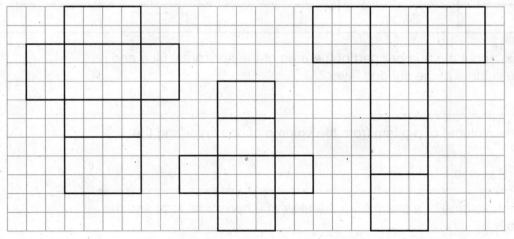

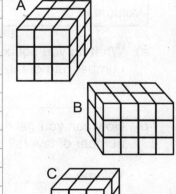

6. Draw a net for the prism on 1 cm grid paper.

a)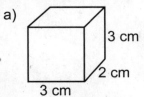

b)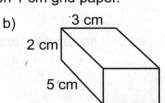

c)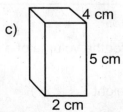

d)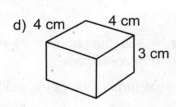

MD5-29 Volume and Area of One Face

1. The top face and the bottom face are the **horizontal faces**. Shade the horizontal faces on the prism.

   a)   b) 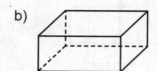   c)   d)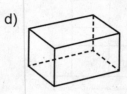

2. The bottom face of the rectangular prism has an area of 12 cm².

   What is the area of the shaded face? _____

   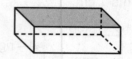

3. These prisms are made from 1 cm cubes. Fill in the table.

Area of a horizontal face	6 cm²			
Volume of shaded layer	6 cm³			
Height of prism	4 cm			
Number of horizontal layers	4			
Volume of prism	24 cm³			

   a) What do you notice about the number part of the area of the horizontal face and number part of the volume of the shaded layer?

   _____

   b) How can you get the volume of a prism from the volume of one layer and the number of layers?

   Volume = _____

   c) How can you get the volume of a prism from the area of a horizontal face and the height of the prism?

   Volume = _____

4. Circle the part of the formula for the volume of a prism that shows the area of a horizontal face.

   Volume = length × width × height

For rectangular prisms,

volume = length × width × height   OR   volume = area of horizontal face × height.

5. Find the volume. Remember to include the units in your answer.

a)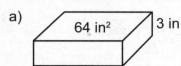

Volume = _____ × _____

= _____

b)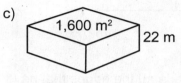

Volume = _____ × _____

= _____

c)

1,600 m² 22 m

Volume = _____ × _____

= _____

d) Area of top face = 25 m²

Height = 10 m

Volume = _____ × _____

= _____

e) Area of top face = 200 mm²

Height = 45 mm

Volume = _____ × _____

= _____

f) Area of top face = 15 m²

Height = 32 m

Volume = _____ × _____

= _____

6. a) Circle the part of the formula for the volume of a prism that shows the area of the shaded face. Volume = length × width × height.

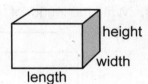

height
width
length

b) Find the volume.

i)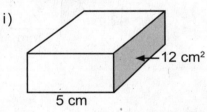

12 cm²
5 cm

Volume = _____

ii)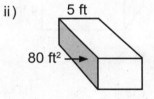

5 ft
80 ft²

Volume = _____

BONUS ▶

3 m
15 m²

Volume = _____

7. a) Tom thinks he can find the volume of a rectangular prism using the formula volume = width × length × height. Is he correct? Explain.

b) Tom thinks the volume of this prism is 30 in² × 6 in = 180 in³. Is he correct? Explain.

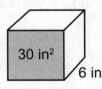

30 in²
6 in

c) Find the volume.

i)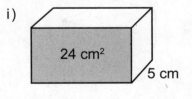

24 cm²
5 cm

Volume = _____

ii)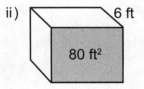

6 ft
80 ft²

Volume = _____

iii)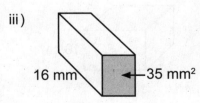

16 mm
35 mm²

Volume = _____

1. Find the missing measurement.

a)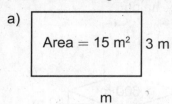

Area = 15 m² | 3 m

_____ m

b)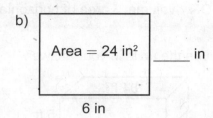

Area = 24 in² | _____ in

6 in

c)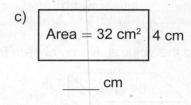

Area = 32 cm² | 4 cm

_____ cm

2. Shade **all** the edges that have the same measurement as the edge marked.

Example: 2 cm 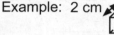 → 2 cm

a)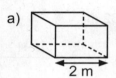

2 m

b)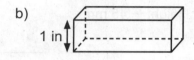

1 in

c)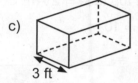

3 ft

3. Write the length of the thick edge beside it. Then find the missing measurement.

a)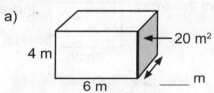

4 m

6 m

20 m²

_____ m

b)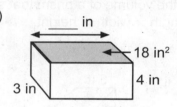

_____ in

3 in

18 in²

4 in

c)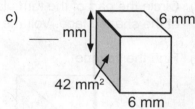

_____ mm

6 mm

42 mm²

6 mm

> **REMINDER ▶** For rectangular prisms,
>
> volume = length × width × height   OR   volume = area of horizontal face × height.

4. Find the volume of the prisms in Question 3.

a) Volume = _____

b) Volume = _____

c) Volume = _____

5. Find the area of the shaded face.

a) Volume = 250 cm³

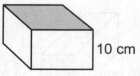

10 cm

Area = _____

b) Volume = 70 m³

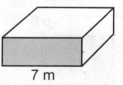

7 m

Area = _____

c) Volume = 48 in³

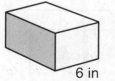

6 in

Area = _____

6. Find the missing measurement.

a) Volume = 36 km³

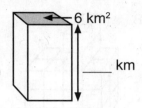

6 km²

_____ km

b) Volume = 63 ft³

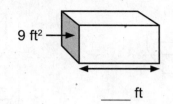

9 ft²

_____ ft

c) Volume = 30 yd³

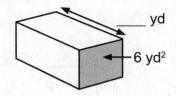

_____ yd

6 yd²

d) Volume = 36 m³

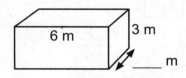

6 m

3 m

_____ m

e) Volume = 1,200 mm³

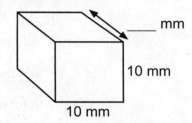

_____ mm

10 mm

10 mm

f) Volume = 378 in³

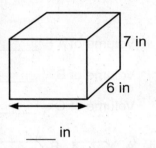

7 in

6 in

_____ in

REMINDER ▶ Area of a horizontal face = length × width.

---

7. The bottom face is labeled on the net. Label and record the length, width, and height of the prism. Then find the volume of the prism.

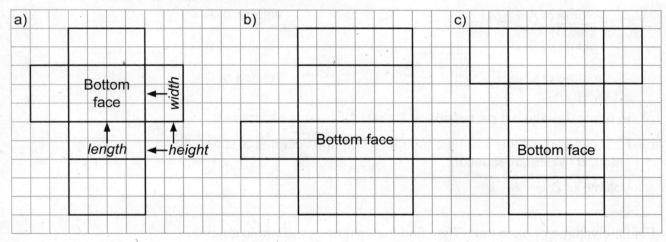

a) Bottom face — width, length, height

b) Bottom face

c) Bottom face

Length = __4__ units

Width = __3__ units

Height = __2__ units

Volume = ____4 × 3 × 2____

= __24__ units³

Length = _____ units

Width = _____ units

Height = _____ units

Volume = _____

= _____ units³

Length = _____ units

Width = _____ units

Height = _____ units

Volume = _____

= _____ units³

MD5-31 Volume of Shapes Made from Rectangular Prisms

1. a) Find the volume of each shape made from 1 cm cubes.

i) A B

C

ii) A B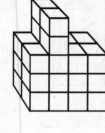

C

Volume of A = _____

Volume of B = _____

Volume of C = _____

Volume of A = _____

Volume of B = _____

Volume of C = _____

iii) A B C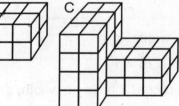

iv) A B C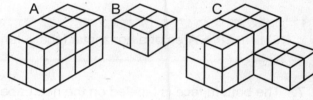

Volume of A = _____

Volume of B = _____

Volume of C = _____

Volume of A = _____

Volume of B = _____

Volume of C = _____

b) In part a), shade the part of shape C that is the same as shape B.

c) How can you get the volume of shape C from the volumes of shape A and shape B? Write an equation.

Volume of C = _____

2. Find the volume of each part. Then find the volume of the whole shape.

a)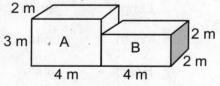

b)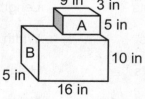

Volume of A = _____

Volume of B = _____

Total volume = _____

Volume of A = _____

Volume of B = _____

Total volume = _____

3. a) A building is 9 stories high. The wing of the building is 5 stories high. How many stories taller is the tower than the wing?

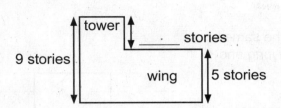

b) The tower of a building is 10 m wide. The base is 60 m wide. How wide is the wing?

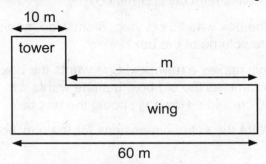

c) What is the total height of the building?

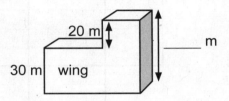

d) What is the width of the tower?

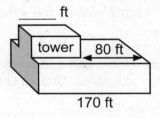

4. Trace all visible edges that are the same measurement as the thick edge.

a)
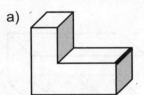

b)
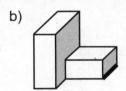

c)
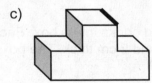

d)
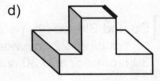

5. Show how to divide the whole shape into rectangular prisms. Use the volume of each prism to find the total volume of the shape.

a)
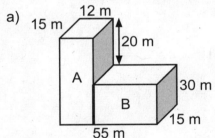

Volume of A = _____

Volume of B = _____

Total volume = _____

b)
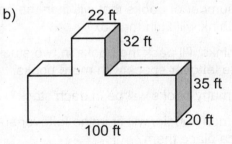

Volume of A = _____

Volume of B = _____

Total volume = _____

c)
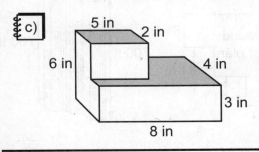

BONUS ▶
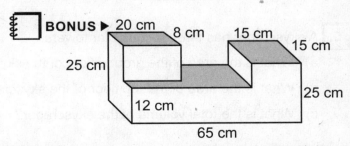

MD5-32 Problems with Volume

1. Ava breaks her clay charms box.

    a) The box was 15 cm long, 8 cm wide, and 9 cm tall. What was the volume of the box?

    b) Ava makes a new box. She wants the box to have the same volume as the old box, but she wants it to be 12 cm long and 10 cm wide. How tall should the box be?

    c) Write the set of dimensions for the new box.

2. Anwar draws a net for a box without a lid. What are the length, width, and height of his box? What is the volume of his box?

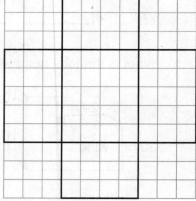

3. Safety deposit boxes in a bank come in two sizes. Anna needs to store about 360 in³ of valuables. Should she choose Box A or Box B?

    Box A: 3 in × 6 in × 24 in   Box B: 3 in × 8 in × 12 in

4. Bo and Blanca are packing books into a box. Each book is 28 cm long, 22 cm wide, and 2 cm thick. The box is 50 cm long, 30 cm wide, and 30 cm tall.

    a) What is the volume of each book?

    b) What is the volume of the box?

    c) Bo divides the volume of the box by the volume of 1 book to find the number of books that will fit in the box. How many books does he think will fit in the box?

    d) Blanca thinks: I'll pack the books in two stacks, as shown. Then I can fill the leftover space with more books.

        i) How many books will be in each stack?

        ii) How many books will fit in the leftover space? How should Blanca place them?

        iii) How many books in total will fit in the box?

    e) Why can Bo and Blanca fit a different number of books in the box? How many books will really fit in the box?

5. A skyscraper has three rectangular towers.

    a) What is the area of the ground floor of the skyscraper?

    b) What is the area of the top floor of the skyscraper?

    c) What is the total volume of the skyscraper?

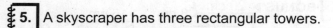

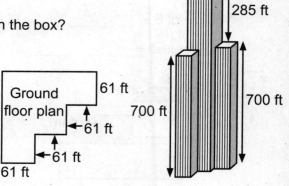

REMINDER ▶ 1 m = 100 cm   1 cm = 10 mm

6. a) Sofia says that the volume of this box is 1 × 0.5 × 80 = 40 cm³. Is she correct? Explain why or why not.

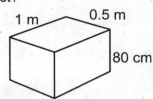

1 m   0.5 m

80 cm

_____

_____

b) Change the measurements of the box into centimeters. Then find the volume.

Volume = _____cm × _____ cm × _____ cm = _____ cm³

7. Find the volume.

a)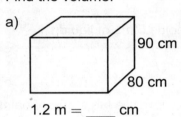

90 cm

80 cm

1.2 m = _____ cm

Volume = _____

b) ERASER

18 mm

12 mm

4.1 cm = _____ mm

Volume = _____

c)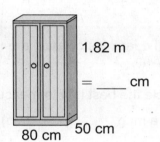

1.82 m

= _____ cm

80 cm   50 cm

Volume = _____

REMINDER ▶ 1 yd = 3 ft   1 ft = 12 in

8. Find the volume.

a)

1 ft

$2\frac{1}{3}$ yd

2 ft

= _____ ft

Volume = _____

b)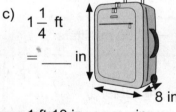

8 in

9 in

2 ft

= _____ in

Volume = _____

c) $1\frac{1}{4}$ ft

= _____ in

8 in

1 ft 10 in = _____ in

Volume = _____

9. A dog travel crate is 1.24 m long, 77 cm wide, and 83 cm tall. What is the volume of the crate?

10. The inside of a microwave is $1\frac{1}{3}$ ft long, 1 ft 2 in deep, and 9 in tall. What is the inside volume of the microwave?

BONUS ▶ A rectangular prism is 1 km long, 1 m wide, and 1 cm tall. Is its volume more than 1 m³? How do you know? Hint: Convert all the lengths to meters.

Measurement and Data 5-32   129

MD5-33 Liquid Volume

The volume of liquids is often measured in **liters** (L).
One liter is a little more than 4 cups.

1. Circle the objects that can hold less than 1 liter.

  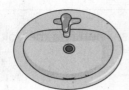  

Small quantities of liquid are measured in **milliliters** (mL). One teaspoon holds 5 mL of liquid.

2. Circle the best unit to measure how much the container can hold.

a) a mug	b) a bathtub	c) a kettle	d) a large can of paint
mL L	mL L	mL L	mL L

3. Circle the amount the container can likely hold.

a) a milk jug	b) a water tank	c) a swimming pool	d) a can of soda
300 mL 300 L	200 mL 200 L	14,000 mL 14,000 L	330 mL 330 L

1 liter = 1,000 milliliters   1 L = 1,000 mL

4. Fill in the table.

L	1	2	3	4	5	6	7	8
mL	1,000							

To change a measurement from liters (L) to milliliters (mL), what number do you

multiply by? _____

5. Convert the measurements in liters to milliliters.

a) 9 L = _____ mL

b) 12 L = _____ mL

c) 40 L = _____ mL

d) 35 L = _____ mL

e) 132 L = _____ mL

f) 200 L = _____ mL

6. Convert the measurement in liters to milliliters. Then circle the greater measurement.

a) 700 mL   3 L   b) 6,300 mL   8 L   c) 23,567 mL   23 L

                    3,000 mL

d) 6,666 mL   6 L   e) 70 L   7,800 mL   f) 75 L   65,203 mL

7. a) Write a measurement in milliliters that is between 7 L and 8 L. _____

    b) Write a measurement in whole liters that is between 6,905 mL and 7,603 mL. _____

> If 1,000 milliliters = 1 liter, then 1 mL = one thousandth of 1 L.
>
> If 1,000 mL = 1 L, then 1 mL = 0.001 L.

8. Fill in the measurements in liters.

mL	1	2	3	4	5	6	7	8	9	10
L	0.001									

9. Convert the measurement in milliliters to liters.

a) 345 mL = __345__ thousandths of 1 L   b) 463 mL = _____ thousandths of 1 L

        = __0.345__ L   = _____ L

c) 81 mL = _____ thousandths of 1 L   d) 40 mL = _____ thousandths of 1 L

       = _____ L   = _____ L

e) 760 mL = _____ thousandths of 1 L   f) 3,906 mL = _____ thousandths of 1 L

       = _____ L   = _____ L

10. Convert the measurement from liters to milliliters.

a) 0.2 L = __200__ thousandths of 1 L   b) 0.839 L = _____ thousandths of 1 L

      = __200__ mL   = _____ mL

c) 0.8 L = _____ thousandths of 1 L   d) 0.07 L = _____ thousandths of 1 L

      = _____ mL   = _____ mL

e) 0.004 L = _____ thousandths of 1 L   f) 1.234 L = _____ thousandths of 1 L

      = _____ mL   = _____ mL

MD5-34 Changing Units of Liquid Volume

Josh multiplies by 1,000 to convert 0.64 L to milliliters. First he writes the decimal as thousandths. Then he shifts the decimal point three places to the right: 0.64 L = 0.640 L = 640 mL.

1. Convert the measurement in liters to milliliters.

   a) 1.7 L = __1.700__ L = __1,700__ mL   b) 0.59 L = _____ L = _____ mL

   c) 2.54 L = _____ L = _____ mL   d) 0.02 L = _____ L = _____ mL

   e) 0.004 L = _____ mL   f) 1.759 L = _____ mL

   g) 1.04 L = _____ mL   h) 24.7 L = _____ mL

Camille divides by 1,000 to convert 68 mL to liters. She knows 68 = 68.0, so she shifts the decimal point three places to the left, adding zeros as she needs: 68 mL = 68.0 mL = 0.068 L.

2. Divide by 1,000 to convert the measurement in milliliters to liters.

   a) 5 mL = __0.005__ L   b) 339 mL = _____ L

   c) 9,083 mL = _____ L   d) 54 mL = _____ L

   e) 3 mL = _____ L   f) 18,951 mL = _____ L

   g) 700 mL = _____ L   h) 60 mL = _____ L

3. a) Write a measurement in milliliters that is between 7.89 L and 7.9 L. _____

   b) Write a measurement in liters that is between 6,905 mL and 6,907 mL. _____

4. Convert the measurement in liters to a mixed measurement.

   a) 6.79 L = __6__ L __790__ mL   b) 3.247 L = _____ L _____ mL

   c) 4.027 L = _____ L _____ mL   d) 5.82 L = _____ L _____ mL

   e) 5.008 L = _____ L _____ mL   f) 12.75 L = _____ L _____ mL

   g) 2.7 L = _____ L _____ mL   h) 58.1 L = _____ L _____ mL

5. Convert the measurement in milliliters to a mixed measurement.

   a) 5,130 mL = __5__ L __130__ mL   b) 5,217 mL = _____ L _____ mL

   c) 4,367 mL = _____ L _____ mL   d) 4,081 mL = _____ L _____ mL

   e) 7,006 mL = _____ L _____ mL   f) 44,300 mL = _____ L _____ mL

6. Convert the mixed measurement to a measurement in milliliters.

a) 3 L = ___3,000___ mL

so 3 L 71 mL =

3	0	0	0	mL
+		7	1	mL
3	0	7	1	mL

b) 4 L = _____ mL

so 4 L 510 mL =

				mL
+				mL
				mL

c) 9 L = _____ mL

so 9 L 45 mL =

				mL
+				mL
				mL

d) 2 L = _____ mL

so 2 L 128 mL =

				mL
+				mL
				mL

e) 9 L = _____ mL

so 9 L 50 mL =

				mL
+				mL
				mL

f) 7 L = _____ mL

so 7 L 2 mL =

				mL
+				mL
				mL

7. Ivan uses a mug that holds 300 mL to fill a pot with water. He fills and empties the mug five times to fill the pot. How many liters can the pot hold?

8. How many containers of the given size are needed to make 1 L?

a) 250 mL   b) 100 mL   c) 200 mL

9. Mona makes fruit punch using 960 mL of apple juice, 360 mL of cranberry juice, 240 mL of orange juice, and 1.5 L of ginger ale. How much fruit punch in total did she make? Write the answer three ways: in milliliters, in liters, and as a mixed measurement.

10. A cafeteria sells coffee in three sizes: small, 240 mL; medium, 350 mL; and large, 470 mL. The cafeteria sells 500 large cups, 1,000 medium cups, and 750 large cups of coffee. How many liters of coffee did the cafeteria sell in total?

11. A 1.89 L bottle of tomato juice costs $3.99. A pack of six cans is on sale for $1.99. Each can is 163 mL.

a) Which contains more juice, one 1.89 L bottle, or two packs of six 163 mL cans?

b) What costs more, one 1.89 L bottle, or two packs of six 163 mL cans?

c) Which way of buying the juice is cheaper by volume? Explain.

MD5-35 Filling Containers

> **Capacity** is the amount of liquid (or rice, beans, and so on) that a container can hold.
>
> A container with a volume of 1 cubic decimeter (1 dm³) has a capacity of 1 liter (1 L).
> A container with a volume of 1 cubic centimeter (1 cm³) has a capacity of 1 milliliter (1 mL).

1. a) What is the volume of a cube with sides 1 dm = 10 cm?

   Volume = 1 dm³ = _____ cm × _____ cm × _____ cm = _____ cm³

   b) 1 dm³ = _____ cm³ and 1 L = _____ mL, so 1 cm³ = _____ mL

2. a) A can has a volume of 4 dm³. What is its capacity? _____

   b) A jar has a volume of 450 cm³. What is its capacity? _____

   c) A can has a capacity of 330 mL. What is its volume? _____

   d) A juice carton has a capacity of 1.89 L. What is its volume? _____

3. Find the volume and the capacity of the aquarium. Include the units!

   a)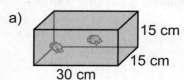

   Length = ____30 cm____

   Width = ____15 cm____

   Height = ____15 cm____

   Volume = ____30 cm × 15 cm × 15 cm____

           = ____6,750 cm³____

   Capacity = ____6,750 mL____

   b) 20 cm / 20 cm / 37 cm

   Length = _____

   Width = _____

   Height = _____

   Volume = _____

           = _____

   Capacity = _____

   c)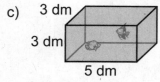
   3 dm / 3 dm / 5 dm

   Length = _____

   Width = _____

   Height = _____

   Volume = _____

           = _____

   Capacity = _____

   d)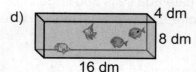
   4 dm / 8 dm / 16 dm

   Length = _____

   Width = _____

   Height = _____

   Volume = _____

           = _____

   Capacity = _____

4. Use two ways to find the capacity of the prism.

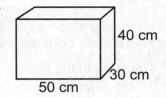

a) Find the volume in cubic centimeters. Then convert cm³ to mL.

Volume = _____ cm³, so capacity = _____ mL.

b) Convert the measurements to decimeters.

Length = _____ dm   Width = _____ dm   Height = _____ dm

c) Find the volume in cubic decimeters. Then convert dm³ to liters.

Volume = _____ dm³, so capacity = _____ L.

d) How should your answers in parts a) and c) be related? Explain. Are they related this way? If not, find your mistake.

_____

_____

5. Find the capacity of the prism.

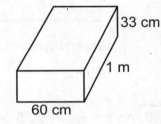

a) Convert all the measurements to centimeters.

Length = _____ cm   Width = _____ cm   Height = _____ cm

b) Find the volume in cubic centimeters. Then convert cm³ to mL.

Volume = _____ cm³, so capacity = _____ mL.

c) What is the capacity of the prism in liters? _____ L

6. An aquarium has a length of 35 cm and a width of 25 cm. The water in the aquarium is 16 cm high. How much water is in the aquarium?

_____ mL = _____ L

> **REMINDER** ▶ Volume of rectangular prism = area of horizontal face × height

7. Find the volume of the prism. Then find the height.

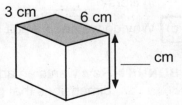

a) Capacity = 36 mL

Volume = _____

b) Capacity = 320 L

Volume = _____

c) Capacity = 90 mL

Volume = _____

When you place an object into water, the level of water rises.
This is called **displacing** the water.

The cube in the picture has sides that measure 1 dm or
10 cm each. The cube has a volume of 1 cubic decimeter.

The cube displaces 1 L of water.

A centimeter cube has a volume of 1 cm³. It displaces 1 mL of liquid.

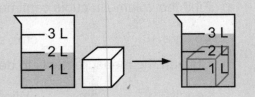

8. A measuring cup contains 50 mL of water. Ron places three 1 cm cubes into it.

   What is the level of the water after Ron places the cubes? _____

9. The water jug contains 400 mL of water and a toy. Find the volume of the toy.

   a)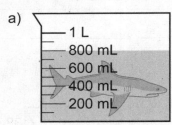

   1 L
   800 mL
   600 mL
   400 mL
   200 mL

   b)

   1 L
   800 mL
   600 mL
   400 mL
   200 mL

   c)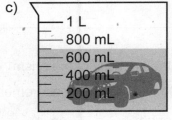

   1 L
   800 mL
   600 mL
   400 mL
   200 mL

   Water displaced: _____ mL   Water displaced: _____ mL   Water displaced: _____ mL

   Volume = _____ cm³   Volume = _____   Volume = _____

10. An aquarium is 25 cm wide and 40 cm long.

   a) Mona pours 8 L of water into the aquarium. How high is the water? Hint: Find the
      volume of the water in cubic centimeters first.

   b) Mona places a stone in the aquarium. The water level rises to 9 cm. What is the
      volume of the stone and the water together?

   c) What is the volume of the stone?

11. The volume of very large amounts of liquid is measured in cubic meters.

   a) 1 m³ = _____ cm × _____ cm × _____ cm = _____ cm³

      How many liters are in 1 m³? _____

   b) A community swimming pool has a volume of 1,800 m³. What is the volume

      in liters? _____

   c) Why do we need a unit between 1 m³ and 1 cm³?

   BONUS ▶ Ravi wants an aquarium that will hold 24 L of water. Write a set of
           dimensions that give the required capacity.

   Length = _____ cm   Width = _____ cm   Height = _____ cm

MD5-36 Cups and Fluid Ounces

> In recipes, amounts of liquid, flour, sugar, and many other ingredients are often measured in **cups** (c).

1. Will a 4-cup bowl be enough for all the ingredients in the recipe?

   a) **Pancakes**

      $1\frac{1}{2}$ cups milk

      2 cups flour

      $\frac{1}{4}$ cup melted butter

      2 eggs $= \frac{7}{16}$ cup of raw eggs

      Total:

      Less than 4 cups? _____

   b) **Blueberry Frozen Yogurt**

      $1\frac{1}{2}$ cups blueberries

      $\frac{3}{8}$ cups sugar

      $\frac{3}{4}$ cups yogurt

      $\frac{1}{4}$ cup milk

      Total:

      Less than 4 cups? _____

> We measure small amounts of liquid in **fluid ounces** (fl oz). There are 8 fluid ounces in 1 cup.

2. Fill in the table.

c	1	2	3	4	5	6	7	8
fl oz	8							

   To convert from cups to fluid ounces, what number do you multiply by? _____

3. Convert the measurement in cups to fluid ounces.

   a) 10 c = _____ fl oz   b) 12 c = _____ fl oz   c) 25 c = _____ fl oz

4. Convert the measurement in cups to fluid ounces. Then circle the larger measurement.

   a) 15 c   100 fl oz   b) 160 fl oz   10 c   c) 50 c   500 fl oz

BONUS ▶

a) Write a measurement in fluid ounces that is between 7 c and 8 c. _____

b) Write a measurement in whole cups that is between 50 fl oz and 60 fl oz. _____

5. Multiply by 8 to convert the measurement in cups to fluid ounces. Include the units.

a) $\dfrac{1}{4}$ c $= \dfrac{1}{4} \times 8 = 1 \times (8 \div 4)$ b) $\dfrac{5}{8}$ c $=$ c) $\dfrac{1}{2}$ c $=$

$= 2$ fl oz

6. Convert the measurement in cups to fluid ounces. Include the units.

a) $1\dfrac{1}{8}$ c $=$ b) $2\dfrac{1}{2}$ c $=$ c) $1\dfrac{3}{4}$ c $=$

There are 8 fluid ounces in 1 cup, so 1 fluid ounce is $\dfrac{1}{8}$ of 1 cup. 1 fl oz $= \dfrac{1}{8}$ c

7. Fill in the table.

fl oz	1	2	3	4	5	6	7	8
c	$\dfrac{1}{8}$							

8. Divide by 8 to convert the measurement in fluid ounces to cups.

a) $45 \div 8 =$ _____ R _____, so b) $67 \div 8 =$ _____ R _____, so c) $90 \div 8 =$ _____ R _____, so

45 fl oz $=$ c   67 fl oz $=$ c   90 fl oz $=$ c

d) $5 \div 8 =$ _____ R _____, so e) $6 \div 8 =$ _____ R _____, so f) $12 \div 8 =$ _____ R _____, so

5 fl oz $=$ c   6 fl oz $=$ c   12 fl oz $=$ c.

9. Kim mixes 2 fl oz of yogurt, 2 fl oz of apple juice, $\dfrac{1}{2}$ c of berries, and $\dfrac{1}{2}$ c of ice to make a serving of a berry smoothie. What is the size of the serving?

2 cups make 1 **pint**. 2 c = 1 pt

1. Multiply by 2 to convert from pints to cups.

a) 3 pt = ___6___ c

b) 5 pt = _____ c

c) 12 pt = _____ c

d) $\frac{1}{2}$ pt = _____ c

e) $5\frac{1}{2}$ pt = _____ c

f) $10\frac{1}{2}$ pt = _____ c

2. Divide by 2 to convert from cups to pints.

a) 8 c = ___4___ pt

b) 12 c = _____ pt

c) 30 c = _____ pt

d) 5 c = _____ pt

e) 9 c = _____ pt

f) 45 c = _____ pt

1 **quart** equals 2 pints or 4 cups. 1 qt = 2 pt = 4 c

3. Multiply by 2 to convert from quarts to pints and from pints to cups.

a) 2 qt = ___4___ pt = ___8___ c

b) 3 qt = _____ pt = _____ c

c) 6 qt = _____ pt = _____ c

d) $\frac{1}{2}$ qt = _____ pt = _____ c

e) $3\frac{1}{2}$ qt = _____ pt = _____ c

f) $10\frac{1}{2}$ qt = _____ pt = _____ c

4. Divide by 2 to convert from pints to quarts.

a) 10 pt = ___5___ qt

b) 25 pt = _____ qt

c) 15 pt = _____ qt

Large volumes of liquid or capacities are measured in **gallons** (gal). 1 gal = 4 qt = 8 pt = 16 c

5. Fill in the table.

gal	1	2	3	4	5	6	7
qt	4						
pt	8						
c	16						

6. Remember: 1 c = 8 fl oz. What number do you multiply by to convert…

a) from gallons to cups? _____

b) from gallons to fluid ounces? _____

7. Convert.

a) 2 gal = _____ c

b) 3 gal = _____ fl oz

c) 12 gal = _____ c

d) 10 gal = _____ fl oz

REMINDER ▶ $\dfrac{3}{8} \times 16 = 3 \times (16 \div 8)$

$= 3 \times 2 = 6$

$2\dfrac{3}{4} \times 16 = \overset{2 \times 4 + 3}{\dfrac{11}{4}} \times 16 = 11 \times (16 \div 4)$

$= 11 \times 4 = 44$

---

8. Convert from gallons to cups.

a) $\dfrac{3}{4}$ gal $= \dfrac{3}{4} \times 16 = 3 \times (16 \div 4)$

$= 3 \times 4 = 12\ c$

b) $\dfrac{5}{8}$ gal $=$

c) $1\dfrac{1}{2}$ gal $=$

d) $2\dfrac{1}{8}$ gal $=$

9. Convert from gallons to fluid ounces.

a) $\dfrac{3}{8}$ gal $= \dfrac{3}{8} \times 128 =$

b) $1\dfrac{1}{4}$ gal $=$

10. Divide by 16 to convert from cups to gallons.

a) 45 c ÷ 16 = __2__ R __13__, so

$45\ c = 2\dfrac{13}{16}\ gal$

b) 67 c ÷ 16 = _____ R _____, so

67 c =

c) 90 c

d) 6 c

11. Order the units from smallest to largest: c, fl oz, gal, pt, qt. _____, _____, _____, _____, _____

Units become smaller as you go down the stairs.

For example, Roy converts 3 pints to fluid ounces.
Roy goes down two steps.

The new unit is $2 \times 8 = 16$ times smaller, so he needs more units.
He multiplies by 16: 3 pt $= 3 \times 16 = 48$ fl oz.

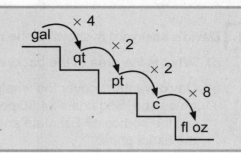

12. a) Convert 12 pt to gallons.

The new units are ___*2 × 4 = 8*___ times

___*larger*___ so I need ___*fewer*___ units.

I _____*divide*_____ by ___*8*___ :

12 ÷ 8 = 1 R 4, so

$12 \text{ pt} = 1\frac{4}{8} = 1\frac{1}{2}$ *gal*

b) Convert 12 qt to cups.

The new units are _____ times

_____ so I need _____ units.

I _____ by _____ :

12 qt =

c) Convert 160 fl oz to quarts.

The new units are _____ times

_____ so I need _____ units.

I _____ by _____ :

160 fl oz =

d) Convert 512 c to gallons.

The new units are _____ times

_____ so I need _____ units.

I _____ by _____ :

512 c =

13. Which holds more juice, a 64 fl oz jar or a $\frac{1}{4}$ gal can? How do you know?

14. A can contains $\frac{3}{4}$ gallons of tomato juice. How many cups does the can contain?

15. There will be 12 guests at Ava's party. Each guest might drink 3 cups of juice.
How many 2 qt cartons of juice should Ava buy?

16. A doctor recommends that Jayden drinks 8 cups of water every day. How many
gallons of water should Jayden drink in a week?

17. A recipe for sauce asks for 2 fl oz of vinegar, $\frac{1}{2}$ c of lemon juice, $\frac{1}{2}$ c of soy sauce,
$\frac{3}{4}$ c of honey, and 3 fl oz of ketchup. How much sauce will the recipe make?

1. David's backyard is shown in the picture.

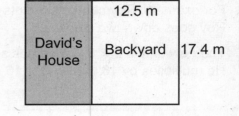

a) What is the area of the backyard?

b) David wants to cover the whole backyard with squares of grass sod. Sod costs $4.30 per square meter. How much will the sod cost? Estimate and then use a calculator to find the exact price.

c) David needs to fence the backyard on three sides. How much fencing does he need?

d) Fencing costs $28 for each meter length of fence. How much will the fencing cost?

e) How much will the fencing and the sod cost together?

2. a) What is the height of the tower?

b) Find the volume of the building.

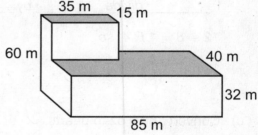

3. Alex wants to make a rectangular flowerbed surrounded by 12 ft of fencing. What length and width of the flowerbed will give him the greatest area?

4. Anika makes a box for her shell collection. The net for the box is shown. Each square on the grid is 1 in long and wide. What is the volume of the box?

5. John opens a new $\frac{1}{2}$ gal jug of milk. He uses 12 fl oz of milk for pancakes and $2\frac{1}{2}$ c for a milkshake. How much milk is left in the jug?

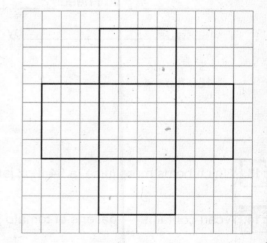

6. An aquarium is 50 cm long and 30 cm wide. The water is 25 cm deep.

a) What is the volume of the water in the aquarium in cubic centimeters? In milliliters? (Remember: $1 \text{ cm}^3 = 1 \text{ mL}$)

b) 1 L of water weighs 1 kg. The empty aquarium weighs 10.5 kg. How much does the aquarium weigh when filled with the water?

c) Ron wants to move the aquarium. Should he move the aquarium when it is full or empty? Explain.

A container with a volume of 231 in³ has a capacity of 1 gallon. 1 gal = 231 in³

7. a) Is the capacity of a 1 in cube more than or less than 1 fl oz? How do you know? Explain without calculating. Hint: How many fluid ounces are in 1 gal?

b) How many cubic inches are in 1 cubic foot?

c) Is the capacity of a 1 ft cube more than or less than 1 gal? How do you know? Explain without calculating.

8. An aquarium is 2 ft long, 14 in wide, and 22 in tall. What is the capacity of the aquarium?

9. An aquarium is 1.2 m long, 60 cm wide, and 85 cm tall.

a) What is the capacity of the aquarium?

b) Maria fills the aquarium with water, equipment, and decorations. The total volume of the contents is 540 L. What height does the water reach?

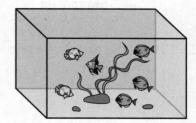

10. A lawn is a rectangle 11 ft wide and 28 ft long.

a) What is the area of the lawn?

b) If 1 ft = 12 in, how many square inches are in 1 square foot? How do you know?

c) What is the area of the lawn in square inches?

d) Imagine a prism that is 1 in tall and its bottom face is the lawn. What is the volume of the prism?

e) In a week, the lawn needs an amount of water equal to the capacity of the prism in part d). How many gallons of water does the lawn need in a week?

11. A café sells one hundred 8 fl oz cups of coffee and two hundred 12 oz cups of coffee.

a) How much coffee in total did the café sell?

b) A small cup of coffee costs $1.29 and a large cup costs $1.49. How much money was the café paid for all the coffee it sold?

c) It costs about $2.20 to make a gallon of coffee. How much would it cost to make the amount of coffee that was sold?

d) How much profit did the café make from coffee?

G5-6 Angles

line segment   ray   line

endpoint   not an endpoint

You can extend a line or a ray as much as needed on the side that has no endpoint.

1. Identify the picture as a line, line segment, or ray. Then write the number of endpoints.

a)

_____

_____ endpoints

b)

_____

_____ endpoints

c)

_____

_____ endpoints

2. Extend the ray at one end.

a)

b)

c)

When 2 rays have the same endpoint, they form an **angle**.
You do not have to draw a dot to show the common endpoint.

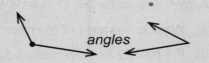 angles

3. Circle the pictures that show angles.

An angle has a **vertex** and **arms**. You can extend the arms
as much as needed without changing the angle.

vertex → 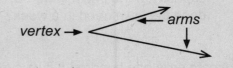 arms

4. Circle the vertex and extend the arms of each angle.

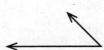   

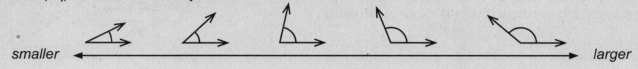

The **size** of an angle is how much you need to turn one arm to get to the other arm.
The **arc** (⌒) shows how much you need to turn.

smaller ⟵————————————————————————⟶ *larger*

5. Circle the larger angle in the pair.

a)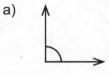

b)

c)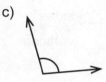

d)

These angles are the same size—you need to turn the *same* amount
to get from one arm to the other arm of both angles. One angle looks
larger because the rays extend farther.

6. Are the angles the same? Extend the arms that look shorter to help you decide.

a)   _____

b)  _____

7. Circle the larger angle. The extend the arms that are shorter to check your answer.

a)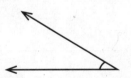

b)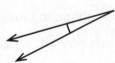

Angles equal to the corners of a square are called **right angles**. You can use
a corner of a sheet of paper to compare an angle with a right angle.

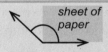

sheet of paper

This angle is greater than a right angle.

8. Compare each angle to a right angle using the corner of a sheet of paper.
Mark the angle as *less* than a right angle or *greater* than a right angle.

a)

_____

b)

_____

c)

_____

d)

_____

G5-7 Sides and Vertices of 2-D Shapes

All two-dimensional (2-D, or flat) shapes have **sides** (or edges) and **vertices** (the endpoints of the sides).

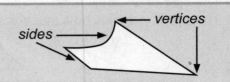

1. Count the number of sides and vertices for the shape. Mark each side with a check and each vertex with a circle as you count.

a)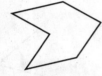

   _4_ sides

   _4_ vertices

b)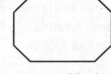

   ____ sides

   ____ vertices

c)

   ____ sides

   ____ vertices

d)

   ____ sides

   ____ vertices

e)

   ____ sides

   ____ vertices

f)

   ____ sides

   ____ vertices

g)

   ____ sides

   ____ vertices

BONUS ▶

   ____ sides

   ____ vertices

A **polygon** is a 2-D shape with sides that are line segments. Sides of polygons do not intersect and 2 sides meet at every vertex. Examples:

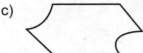

polygons   not polygons

2. Which shape in Question 1 is not a polygon? _____

Explain why this shape is not a polygon. _____

_____

Polygons are named according to how many sides they have.

3 sides: **triangle**   4 sides: **quadrilateral**   5 sides: **pentagon**   6 sides: **hexagon**

3. Count the sides. Then name the polygon.

a)

   _4_ sides

   quadrilateral

b)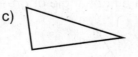

   ____ sides

   _____

c)

   ____ sides

   _____

d)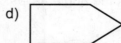

   ____ sides

   _____

4. Complete the chart.

Triangle	Quadrilateral	Pentagon	Hexagon	Other
A,				

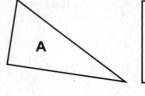

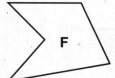 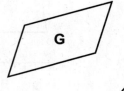 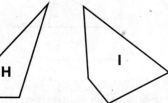 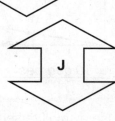

5. Which column in the table in Question 4 does this shape belong to? Explain.

_____

_____

6. Use a ruler to draw the named polygon.

   a) triangle   b) quadrilateral   c) pentagon   d) hexagon

  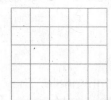 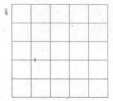

7. a) Count the vertices in the polygons you drew in Question 6. What do you notice?

   b) Can you draw a polygon in which the number of sides *does not* equal the number of vertices?

8. a) Use a ruler. Draw a shape that has the given number of sides and that is *not* a polygon.

   i) 3 sides   ii) 4 sides

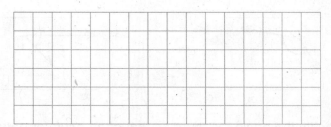

   b) Draw a shape with 1 side and 0 vertices. Is it a polygon?

G5-8 Measuring Angles

We measure angles in **degrees**.

Example: The angle below measures 1 degree.

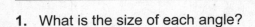

1. What is the size of each angle?

a)

____10 degrees____

b)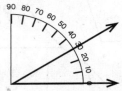

_____

c)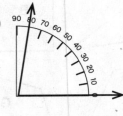

_____

We use a small raised circle after the number instead of the word "degree": 1 degree = 1°.
A right angle measures 90°.

2. Identify the angle as less than 90° or more than 90°.

a)

____less than 90°____

b)

_____

c)

_____

d)

_____

e)

_____

f)

_____

Acute angles are less than a right angle. They measure between 0° and 90°.
Obtuse angles are greater than a right angle. They measure between 90° and 180°.

3. Identify the angle as *acute* or *obtuse*.

a)

_____

b)

_____

c)

_____

d)

_____

e)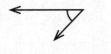

_____

f)

_____

4. Identify the angle measure as that of an *acute* or *obtuse* angle.

a) 55° _____

b) 130° _____

c) 66° _____

d) 93° _____

e) 178° _____

f) 19° _____

To measure an angle, we use a **protractor**.

A protractor has 180 subdivisions of 1° around its curved side.
It has two scales, to measure angles starting from either side.

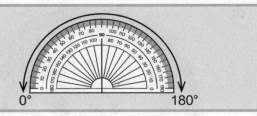

0°   180°

5. Identify the angle as acute or obtuse.
Circle the two numbers that the arm of the angle passes through.
Pick the correct angle measure. (Example: if you said the angle is acute,
pick the number that is less than 90.)

a)

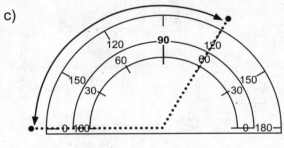

The angle is _____ *acute* _____.

The angle measures ___*60°*___.

b)

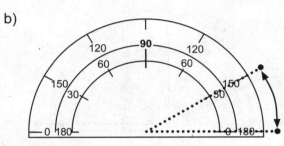

The angle is _____.

The angle measures _____.

c)

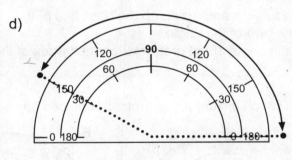

The angle is _____.

The angle measures _____.

d)

The angle is _____.

The angle measures _____.

6. Identify the angle as acute or obtuse. Then write the measure of the angle.

a)

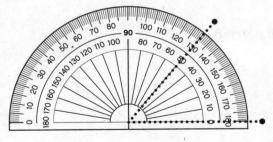

_____

b)

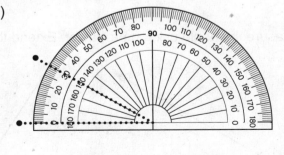

_____

c)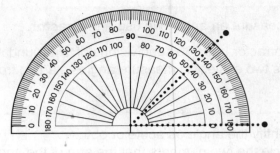

d)

e)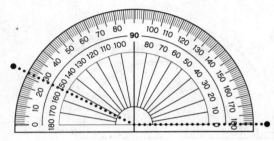

f)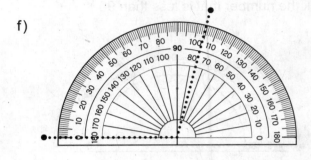

Each protractor has a **base line** and an **origin**.

To measure an angle, line up the base line of the protractor
with one arm of the angle.
Place the origin of the protractor on the vertex of the angle.

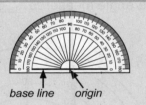

base line   *origin*

7. a) In which picture is the protractor placed correctly? _____

A 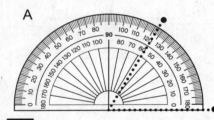   B 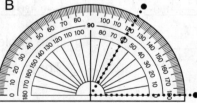   C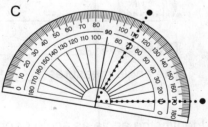

b) What is wrong with the other pictures?

8. Measure the angle using a protractor. Extend the arms if needed.

a) 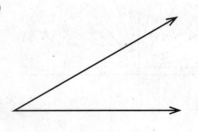   b) 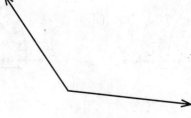   c)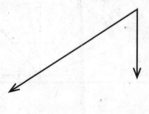

_____   _____   _____

G5-9 Measuring and Constructing Angles

Right angles measure 90°. We mark them with a small square.

Right angles measure between 0° and 90°.
Obtuse angles measure between 90° and 180°.

1. Mark each right angle in the shape with a small square, each acute angle with an *A*, and each obtuse angle with an *O*.

a) 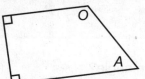   b) 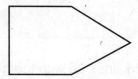   c) 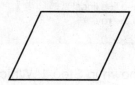   d)

e) 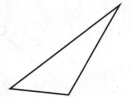   f)   g)    h)

2. Measure the angles in the triangle.

a) 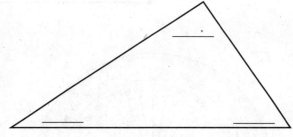   b)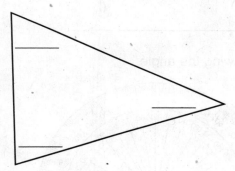

3. Extend the arms of the angle marked with an arc. Then measure the angle.

a) 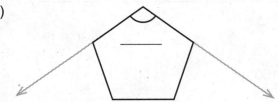   b)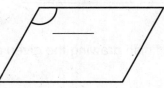

c) 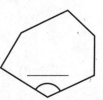   d)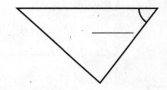

To draw a 60° angle:

Step 1: Draw a ray. Place the protractor as shown.

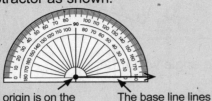

The origin is on the endpoint of the ray.

The base line lines up with the ray.

Step 2: Make a mark at 60°.

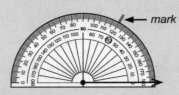

← mark

Step 3: Use a ruler to draw a ray from the endpoint to the mark.

← mark

4. Place the protractor as shown in Step 1. Which mark lines up with the given angle?

a) 60°

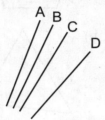

b) 140°

5. Finish drawing the angle.

a) 130°

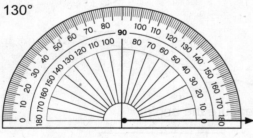

b) 75°

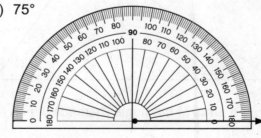

6. Use a protractor to finish drawing the given angle.

a) 40°

b) 160°

7. Use a protractor to draw the angle.

a) 35°   b) 135°   c) 72°   d) 116°

G5-10 Venn Diagrams

Objects that have some property or feature in common make a **group**. We use ovals to show groups. Objects inside the oval have the property. Objects outside the oval do not have the property.

Use these shapes for Questions 1, 2, and 3.

   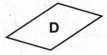

1. Put the letters from each shape inside or outside the circle.

a)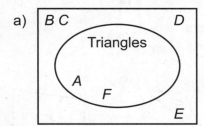

b)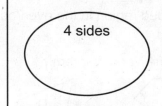

c)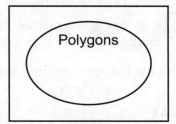

An object can be in more than one group at the same time. We can show this with overlapping ovals, which are called a **Venn diagram**.

2. a) Shade the part where the ovals overlap. Put the letters for shapes in *both* groups in the shaded region.

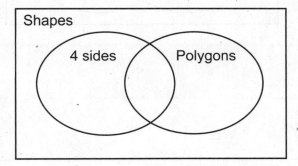

b) Shade the part that is *outside* both ovals. Put the correct letters in that region.

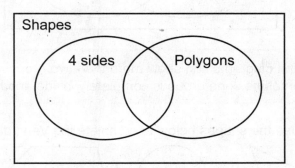

3. Complete the Venn diagrams.

a)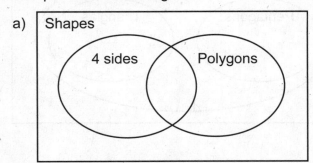

b)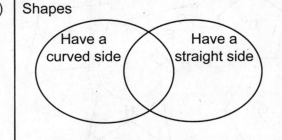

4. a) Complete the Venn diagram using $2\frac{1}{3}$, 3, 7, 12.

i)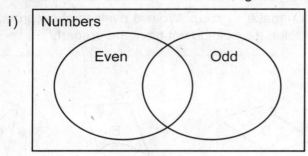

ii)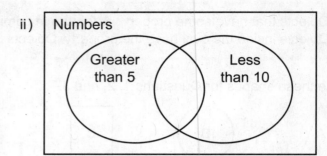

b) In each Venn diagram above, shade the region that stayed empty.

c) Add a number to all non-shaded regions.

d) Can you add a number to the shaded region in the even/odd numbers diagram above? Explain.

5. Use the polygons below to complete the Venn diagram.

a)

b)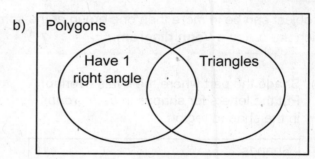

Venn diagrams can show more than two groups.
Sometimes one group is completely inside another group.

---

6. Use the shapes below to complete the Venn diagram.

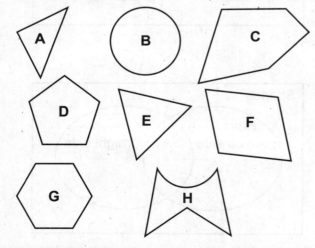 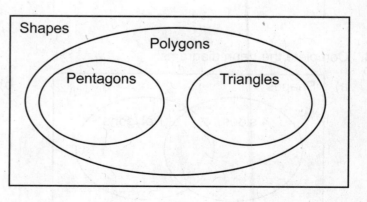

G5-11 Classifying Polygons

1. Measure the sides of the polygon to the closest millimeter. Mark equal sides
   with hash marks.

   a)
   _____ mm

   _____ mm _____ mm

   _____ mm

   b)
   _____ mm

   _____ mm _____ mm

   _____ mm

A polygon that has *all sides* the same length is called an **equilateral polygon**.

2. Measure the sides of the polygon to the closest millimeter.
   Color the equilateral polygons.

   a)
   _____ mm

   _____ mm

   _____ mm

   b)
   _____ mm

   _____ mm

   _____ mm

   _____ mm

   _____ mm

   c)
   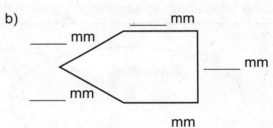
   _____ mm

   _____ mm

   _____ mm

   _____ mm

   d)
   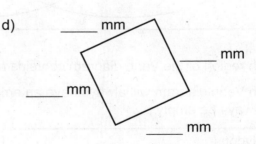
   _____ mm

   _____ mm

   _____ mm

   _____ mm

3. Measure the angles of the polygon. Color the polygons with all equal angles.

   a)
   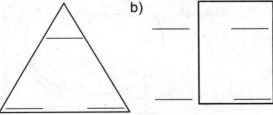

   b)

   c)
   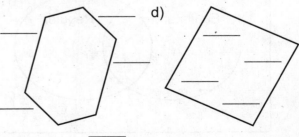

   d)

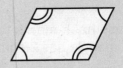

Regular polygons are equilateral and have all angles the same size:

To show angles are equal, you can mark them with arcs,
double arcs, or (for right angles) small squares.

4. a) Circle the regular polygons.

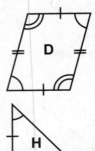

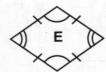
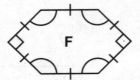

b) Sort the polygons. Some polygons will be in both rows.

Polygons with all angles equal	A,
Equilateral polygons	A,

c) Use the table above to sort the polygons in the Venn diagram.

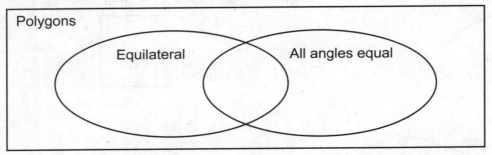

d) Which region of the Venn diagram contains all regular polygons? _____

e) Which Venn diagram will always have an empty region? Shade the region that
   will always be empty.

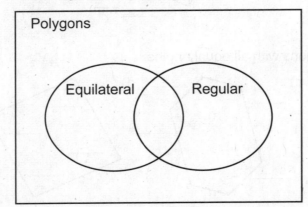
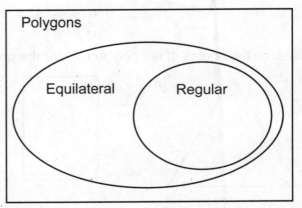

G5-12 Classifying Triangles

1. Measure and label the angles in the triangle. Extend the arms of the angles if needed.

   a)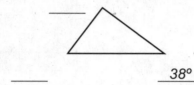
   _____ _____ _38°_

   b)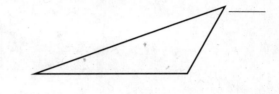
   _____ _____ _____

2. Mark the largest angle in the triangle with an arc.

   a) 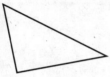   b)    c)   d)

You can classify triangles by the size of the largest angle.

acute-angled or **acute** triangle

The largest angle is acute.

right-angled or **right** triangle

The largest angle is right.

obtuse-angled or **obtuse** triangle

The largest angle is obtuse.

---

3. Mark the largest angle in the triangle with an arc. Use a square corner to check if the angle is acute, right angle, or obtuse. Then classify the triangle.

   a)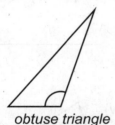
   obtuse triangle

   b)
   _____

   c)
   _____

   d)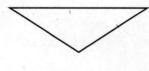
   _____

4. Classify the triangle.

   a)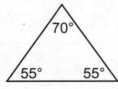
   70°
   55° 55°
   _____

   b)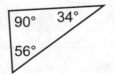
   90° 34°
   56°
   _____

   c)
   60°
   60°
   60°
   _____

   d)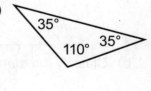
   35°
   110° 35°
   _____

5. Use a ruler to draw the named triangle.

a) acute triangle   b) right triangle   c) obtuse triangle   d) triangle without a right angle

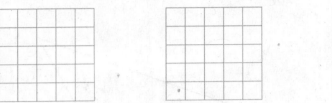

6. How many acute angles does an acute triangle have? Explain. _____

_____

You can classify triangles by the number of *equal* sides.

isosceles triangle

At least 2 equal sides.

scalene triangle

No equal sides.

Isosceles triangles with 3 equal sides are called **equilateral** triangles.

7. Measure the sides of the triangle that seem to be equal to the nearest millimeter.
Mark the equal sides. Then classify the triangle as equilateral, isosceles, or scalene.

a)

_____

b)
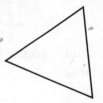
_____

c)

_____

d)
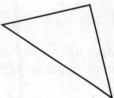
_____

8. a) Classify the triangle as equilateral, isosceles, or scalene.

i)

ii)
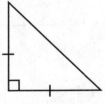

iii)
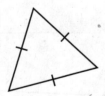

iv)
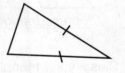

_____   _____   _____   _____

b) Classify the triangles above as acute, right, or obtuse.

i) _____   ii) _____   iii) _____   iv) _____

Geometry 5-12

9. a) Mark the triangle for any equal sides, equal angles, and right angles. Use a ruler to check for equal sides. Use a protractor to check angles.

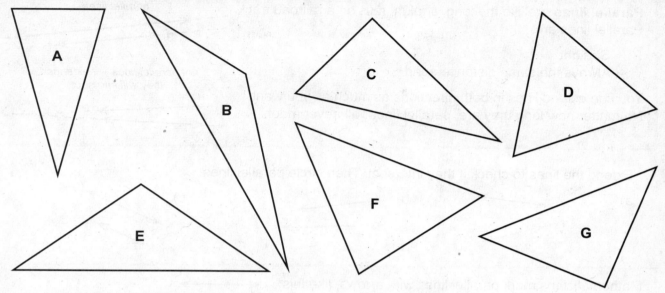

b) Classify the triangles in the tables below.

Acute-angled	
Right-angled	
Obtuse-angled	

Equilateral	
Isosceles	
Scalene	

c) Sort the triangles into 2 groups in a Venn diagram.

i)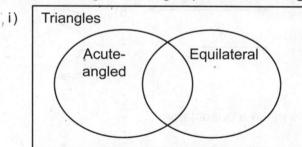

ii)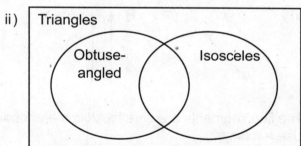

iii) Group 1: right-angled

Group 2: scalene

iv) Group 1: acute-angled

Group 2: scalene

d) Pick 1 group from each table in part b) and make a new Venn diagram. Sort the triangles into 2 groups to complete your Venn diagram.

10. a) In each triangle in Question 9, look at the two angles that are *not the largest*. Are they acute, right, or obtuse angles?

b) Can you draw a triangle with two right angles or two obtuse angles? Explain.

G5-13 Parallel Lines

> **Parallel lines** are like the long, straight rails on a railroad track.
> Parallel lines are:
>
> - Straight
> - Always the same distance apart
>
> You can extend lines in both directions as much as you want.
> No matter how long they are, parallel lines will never meet.

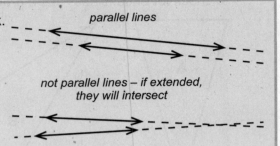

1. Extend the lines to check if they intersect. Then circle parallel lines.

   a)   b)   c)

> Mathematicians mark parallel lines with arrows, like this: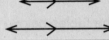

2. Mark any pairs of lines that are parallel with arrows.

   a)    b) 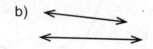   c)    d)

   e) 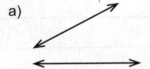   f)   g) 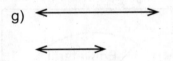   h)

3. The line segments are parallel. Join the endpoints to make a quadrilateral.
   Use a ruler.

   a) 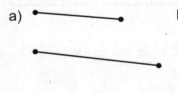   b)   c) 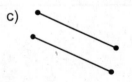   d)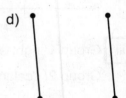

4. Mark the parallel sides with arrows.

   a) 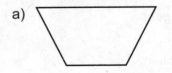   b)   c) 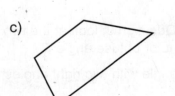   d)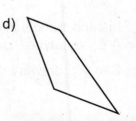

If there is more than one set of parallel sides in a shape,
use a different number of arrows to mark each set.

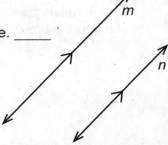

5. Use arrows to mark all the sets of parallel sides. Circle the shapes that
    have no parallel sides.

a)

b)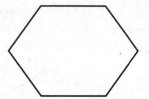

c)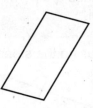

d)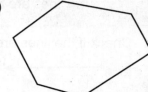

e)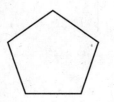

f)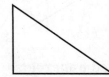

g)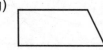

h)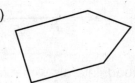

i)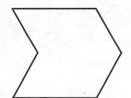

j)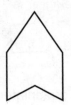

k)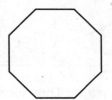

BONUS ▶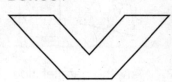

6. a) What is the largest number of **pairs** of parallel sides the polygon can have?
        Draw a polygon with this number of pairs of parallel sides.

    i) quadrilateral _____   ii) hexagon _____   iii) triangle _____   **BONUS** ▶ pentagon _____

    b) On grid paper, draw a quadrilateral and a hexagon that have no parallel sides.

Perpendicular lines are lines that make a right angle.

7. a) Draw a line perpendicular to line *m* that crosses line *n*.

    b) The line you drew in part a) and line *n* make an angle. Measure the angle. _____

    c) Draw a pair of parallel lines by tracing the opposite sides of a ruler.
        Repeat parts a) and b) with the lines. What do you notice?

    d) How can you use right angles to draw parallel lines?

G5-14 Trapezoids and Parallelograms

Follow these steps to check that 2 lines are parallel:

Step 1: Draw a line perpendicular (at a right angle) to 1 of the lines.

Step 2: Measure the angle the new line makes with the second line.

Step 3: If the angle is 90°, the lines are parallel. If it is not a right angle, the lines are not parallel.

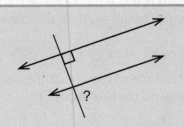

1. Check if the lines are parallel using the steps above. Mark parallel lines with arrows.

a) b)  c)    d)

Quadrilaterals with exactly 1 pair of parallel sides are called **trapezoids**.
Quadrilaterals with 2 pairs of parallel sides are called **parallelograms**.

Trapezoids 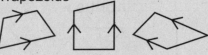   Parallelograms

---

2. Mark the parallel sides with arrows. Then identify the type of quadrilateral.

a)    b)    c) 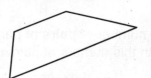   d)

_____   _____   _____   _____

3. a) Sort the quadrilaterals to complete the Venn diagram.

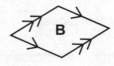   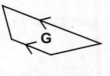

b) Explain why a quadrilateral cannot be a parallelogram and a trapezoid at the same time.

Quadrilaterals

Parallelograms   Trapezoids

                                   A

4. a) Measure the parallel sides of the quadrilaterals to the closest millimeter.

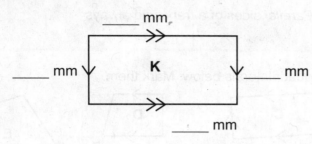

_____ mm

_____ mm **K** _____ mm

_____ mm

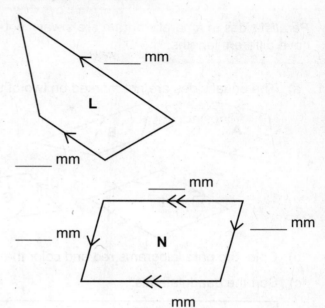

_____ mm

L

_____ mm

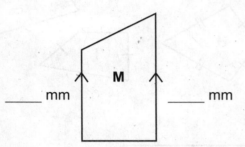

M

_____ mm _____ mm

_____ mm

_____ mm **N** _____ mm

_____ mm

b) Sort the shapes into the tables.

Parallelograms	
Trapezoids	

Parallel sides are equal	
Parallel sides are not equal	

c) What do you notice about the tables? _____

_____

d) One pair of parallel sides is marked. Use your answer in part c) to check if the
other pair of sides is parallel. Then identify the type of quadrilateral.

i)

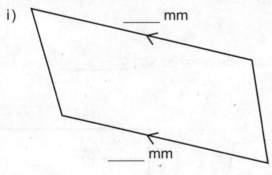

_____ mm

_____ mm

_____

ii)

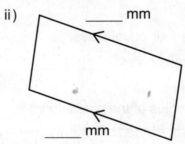

_____ mm

_____ mm

_____

5. Ron thinks that the quadrilateral shown is a parallelogram because
the opposite sides are equal. Is he correct? Explain.

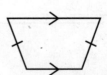

_____

_____

_____

G5-15 Properties of Parallelograms and Trapezoids

Parallel sides in a parallelogram are always equal. *Parallel* sides of a trapezoid always have different lengths.

1. a) The equal sides are not marked on two of the quadrilaterals below. Mark them.

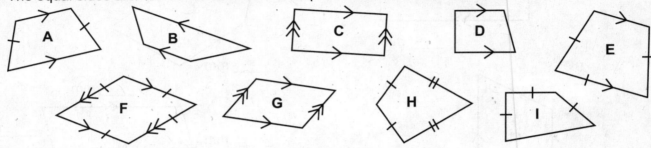

b) Color the parallelograms red and color the trapezoids blue.

c) Sort the quadrilaterals.

No equal sides	
Exactly 2 equal sides	
Exactly 3 equal sides	
4 equal sides (equilateral)	
2 different pairs of equal sides	

d) How many equal sides can a parallelogram have? _____

e) How many equal sides can a trapezoid have? _____

2. A quadrilateral has four equal sides.

a) Can it be a parallelogram? _____   b) Can it be a trapezoid? _____

3. a) Measure the angles of the parallelogram.

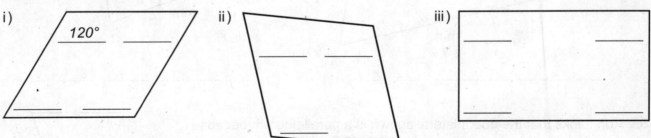

b) What do you notice about the opposite angles in a parallelogram? _____

c) Look at the trapezoids in Question 1. Do they have equal opposite angles? _____

4. Maria drew several trapezoids and parallelograms. Then she covered parts of them with a sheet of paper. Write a "T" on the trapezoids and a "P" on the parallelograms.

A line segment joining 2 opposite vertices of a quadrilateral is called a **diagonal**. Each quadrilateral can have 2 diagonals.

5. a) Draw a different diagonal on each copy of the quadrilateral.

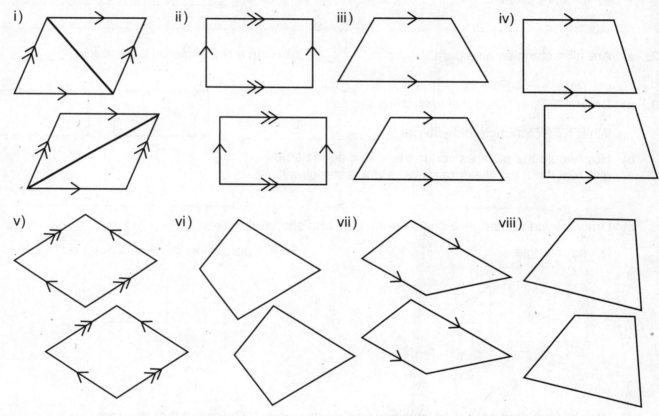

i) ii) iii) iv)

v) vi) vii) viii)

b) A diagonal splits the quadrilateral into 2 triangles. Circle the pictures where these triangles are exactly the same.

c) Look at the quadrilaterals that you circled in part b). Are they parallelograms, trapezoids, or other quadrilaterals? _____

d) Jake put together a right triangle and an obtuse triangle and got a quadrilateral. He thinks his quadrilateral is a parallelogram. Is he correct? Explain.

6. Copy the triangles shown on grid paper and cut them out. Place the 2 triangles in different ways so that they share a side. What types of quadrilaterals can you make this way? Can you make a parallelogram? A trapezoid?

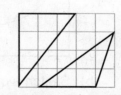

G5-16 Rectangles

A **rectangle** is a quadrilateral with 4 right angles.

1. Use a corner of a sheet of paper to check for right angles. Mark the right angles.
   Circle the rectangles.

   a)   b)   c)   d)   e)   f)

   If two lines both meet a third line at a right angle,
   the two lines are parallel.

   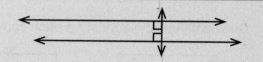

2. a) Are all rectangles also parallelograms? _____   b) Can a rectangle be a trapezoid? _____

3. a) Draw a different diagonal in each rectangle.

      What type of triangle did you get? _____

   b) How would the pictures you made with diagonals be
      different if the parallelograms were *not* rectangles?

      _____

   c) Draw a second triangle with the same size and shape to make…

      i) a rectangle   ii) a parallelogram that is not a rectangle

      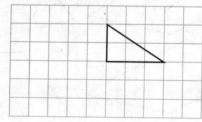   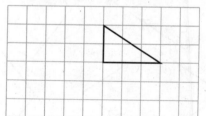

4. a) Which Venn diagram will always have an empty region? Cross out that diagram.

   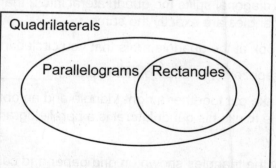

   b) Sketch a quadrilateral in every region of the diagram that is not crossed out.

5. a) Measure the diagonals in the quadrilaterals to the closest millimeter.

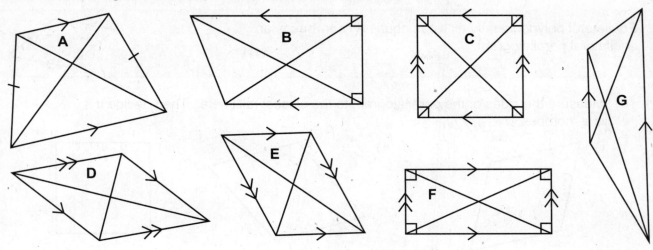

b) Fill in the Venn diagram.

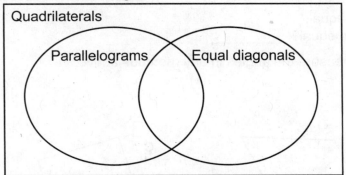

c) Which of the quadrilaterals are rectangles? _____

d) Where are the rectangles in the Venn diagram? _____

e) What do you notice about the diagonals of a rectangle that is not true for diagonals of other parallelograms?

6. a) Explain why all rectangles are also parallelograms.

b) A rectangle has one side 15 cm and another side 6 cm. What are the lengths of the other two sides? Explain. Use the word "parallelogram" in your explanation.

7. The sum of the angles in a quadrilateral is always 360°.

a) Yu thinks that any quadrilateral with 4 equal angles should be a rectangle. Is she correct? Explain.

b) Explain why there cannot be a quadrilateral with exactly 3 right angles.

BONUS ▶ Explain why any parallelogram that has at least 2 right angles has to be a rectangle.

G5-17 Rhombuses

Equilateral polygons have all sides equal. A **rhombus** is an equilateral parallelogram.

1. a) Measure the sides of the parallelogram to the closest millimeter. Then decide if it is a rhombus.

    i) 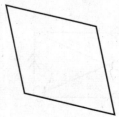   ii) 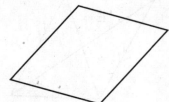   iii)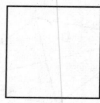

    _____   _____   _____

    b) Opposite angles in a parallelogram are equal.
        Should opposite angles in a rhombus be equal? _____

    c) To check your prediction from part b), measure the angles in any rhombus above.

2. The polygons below are rhombuses.

    A 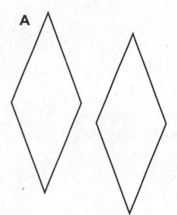   B 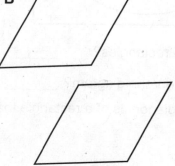   C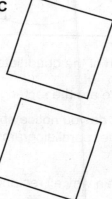

    a) Mark the equal sides with hash marks.

    b) Draw a different diagonal in each copy of the rhombus.

    c) There are two triangles in each picture. Are the two triangles the same or different? _____

    d) Identify the type of triangles in each picture. Write two letters in each triangle:

        i)   E for equilateral *or* I for isosceles *or* S for scalene triangle

        ii) A for acute-angled *or* R for right triangle *or* O for obtuse-angled

    e) What do you notice? How is a rhombus different from other parallelograms?

    f) Measure the diagonals in the rhombuses. Which one of these rhombuses could also be a rectangle? Check your prediction.

3. a) Measure 1 angle between the diagonals in each parallelogram.

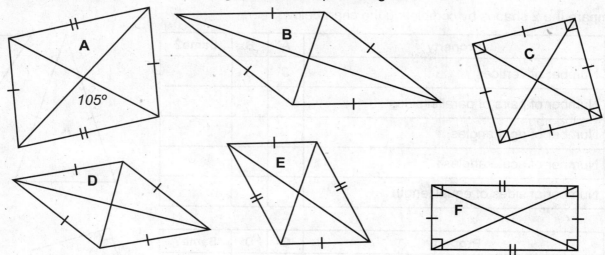

b) Fill in the Venn diagram with letters for the shapes above.

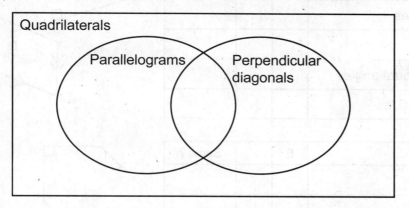

c) Which of the quadrilaterals in part a) are rhombuses? _____

d) Where are the rhombuses in the Venn diagram? _____

e) What do you notice about the diagonals of a rhombus that is not true for diagonals of other parallelograms?

A **square** is an equilateral rectangle, or a rectangle with equal sides

4. a) Is any square also a parallelogram? Explain.

b) Is any square also a rhombus? Explain.

c) Can a square also be a trapezoid? Explain.

d) Shade the region in the Venn diagram where all squares will be.

e) Can there be a polygon that is in the shaded region but is not a square? If yes, sketch it. If no, explain why not.

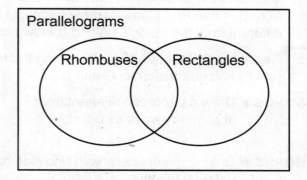

G5-18 Sorting and Classifying Shapes

1. Compare the 2 shapes by completing the chart below.

a)

Property	A	B	Same?
Number of vertices	3	3	
Number of pairs of parallel sides			
Number of right angles			
Number of acute angles			
Number of sides of equal length			

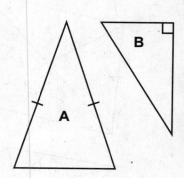

b)

Property	C	D	Same?
Number of sides			
Number of pairs of parallel sides			
Number of right angles			
Number of pairs of opposite equal angles			
Number of pairs equal-length sides			

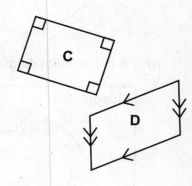

c)

Property	E	F	Same?
Number of vertices			
Number of sides			
Number of pairs of parallel sides			
Number of right angles			
Equilateral			
Regular polygon			

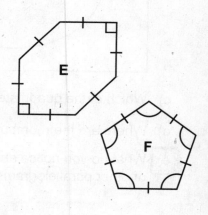

d) Look at the table in part a). Draw a triangle that is different from triangle A in one row and different from B in one row. Can you draw a triangle that is different from the triangles A and B in *two* rows?

e) Repeat part d) for quadrilaterals using the table in part b) and quadrilaterals C and D.

BONUS ▶ Draw a pentagon different from F in as many ways as possible.

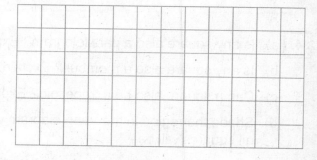

BONUS ▶ Draw a quadrilateral with one side perpendicular to two other sides of different lengths.

2. a) Name the type of quadrilateral. Give the most specific name possible.

A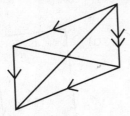

_____

B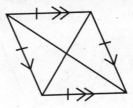

_____

C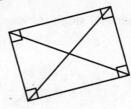

_____

D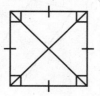

_____

E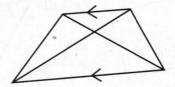

_____

F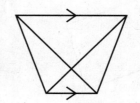

_____

b) Which of the properties does each special quadrilateral from part a) have?
Mark ✓ or ✗.

Quadrilateral	A	B	C	D	E	F
2 pairs of parallel sides						
2 pairs of opposite angles that are equal						
All angles equal						
2 pairs of opposite sides of equal length						
Equilateral polygon						
Diagonals of equal length						
Perpendicular diagonals						

3. a) Sara thinks that the quadrilateral shown is a square because its diagonals
are perpendicular and equal in length. Is she correct? Explain.

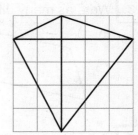

b) On grid paper, sketch another quadrilateral that is not a parallelogram,
but has diagonals that are equal and perpendicular.
Hint: Start with the diagonals.

4. Name the polygons based on the description.

a) a quadrilateral with 4 equal sides and 4 equal angles _____

b) an equilateral quadrilateral with 2 acute angles _____

c) a triangle with 1 right angle and 2 equal sides _____

G5-19 Problems and Puzzles

1. Mark parallel sides, sides of equal lengths, right angles, and equal angles.

a) 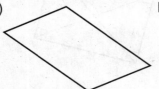   b) 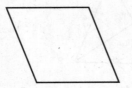   c) 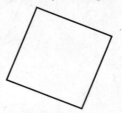   d)

e) 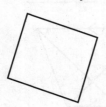   f) 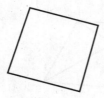   g)    h)

2. Draw a different triangle on each grid. Name the triangle.

a)   b)   c)    d)

_____

_____

e)    f)   g)    h)

_____

_____

3. Write as many different names as you can for each polygon.

a) 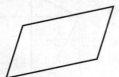   b)   c)   d)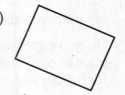

4. On grid paper, sketch two different quadrilaterals with the given property below. Make one quadrilateral of *any type* and try to make the other a *special type* (rectangle, square, parallelogram, rhombus, or trapezoid).

a) exactly 2 right angles

b) perpendicular diagonals

c) equal diagonals

d) exactly 3 equal sides